612.6
Loke,
Life's
astonishing role of the
placenta

LIFE'S VITAL LINK

LIFE'S VITAL LINK

the astonishing role of the placenta

Y. W. LOKE

OXFORD
UNIVERSITY PRESS

OXFORD
UNIVERSITY PRESS

Great Clarendon Street, Oxford, OX2 6DP,
United Kingdom

Oxford University Press is a department of the University of Oxford.
It furthers the University's objective of excellence in research, scholarship,
and education by publishing worldwide. Oxford is a registered trade mark of
Oxford University Press in the UK and in certain other countries

First Edition published in 2013

Impression: 1

British Library Cataloguing in Publication Data
Data available

Library of Congress Cataloging in Publication Data
Data available

ISBN 978-0-19-969451-8

Printed in Great Britain by
Clays Ltd, St Ives plc

For all who have touched my life.
'No man is an island.' John Donne

Author's Notes

Throughout my academic career, all my publications have been directed at fellow scientists. I never attempted to cross the divide from scientific literature to popular science. Now that I have retired from active research, I thought this might be an opportune moment to write for a wider audience, especially as the placenta is a subject that has yet to capture the public's attention. It should be a suitable topic for such a venture.

I decided to approach Oxford University Press because I like the intellectual level at which their popular science volumes are pitched. Not being an established figure in this genre, and without the guidance of an agent, I did not know what kind of response to expect. So I was pleasantly surprised to receive an immediate reply from OUP's Senior Commissioning Editor, Latha Menon: 'The idea of a book on the human placenta is an interesting one. Do send me further details.' This was how a vague idea at the back of my mind became reality. It took exactly nine months from my first letter of enquiry to my signing the contract. For a book on the placenta, I take this to be an auspicious omen. Not only did Latha commission the book, the final manuscript has greatly benefitted from her skilful editing. I owe her a double acknowledgement.

Rotraud Hansberger (philosopher) was the first person to view a few preliminary chapters of the book. I wanted to gauge the reaction of a reader who was not a scientist. Her response was so enthusiastic I was persuaded that the project was viable. Barry Keverne (neuroscientist) kindled my interest in the alliance between the placenta and the brain. Ann Frost (hispanist) kindly polished my grammar. So many people from different areas of reproductive biology have given their valuable time to talk to me that it is not possible to thank them all individually. Their names are featured prominently throughout the book. This should serve as my acknowledgement to them.

Many of the ideas in this book originated from my close research collaboration over many years with Ashley Moffett. While my own interest has been largely focused on the placenta, Ashley's expertise lies with the uterus. Together, we successfully covered both sides of the fetal–maternal relationship.

Sue Griffin devoted many hours to typing the manuscript, right from the beginning when I handed her sheets of seemingly illegible scrawl. Every line in this book represents her fine handiwork. Anne Marie Catchpole helped me compile the endnotes and glossary. All the drawings are by Emily Evans.

This book is for my family. I hope it will give them pleasure.

Contents

1

After the Afterbirth

'The history of man for the nine months preceding his birth would, probably, be far more interesting and contain events of greater moment than all the three-score and ten years that follow it.'   Samuel Taylor Coleridge

The baby within the mother's womb is not alone. It has a constant companion throughout pregnancy. This is the placenta. But the placenta is more than just a companion. It has a very important job to do which is to connect the baby with its mother. This link is crucial because the baby's life-support systems are not yet fully functioning at this early stage of development. Meanwhile, it has to 'borrow' whatever it needs from the mother, using the placenta via the umbilical cord like a jump lead to plug into the mother's battery while its own is flat. The placenta brings in oxygen and food from the mother and, at the same time, excretes waste products back to the mother for her to dispose of. It manufactures products vital for sustaining pregnancy and also forms an effective screen which protects the baby from harmful agents coming from the mother. To perform these complex tasks simultaneously requires flawless organization by the placenta. Even minor disruptions can have serious consequences not only for the baby, but also for the mother.

Yet in spite of being the star of the show, the placenta has never quite managed to gain the attention it deserves. Compared to other organs of the body, the placenta comes very low down the list when measured on the scale of public awareness. It is not the placenta's fault that it has languished in such obscurity. No one has bothered to speak up on its behalf and to put it firmly into the public domain. This book intends to do so. It will trace the life of the placenta from the very moment it is created within a fertilized egg, through its brief but eventful stay within the womb and finally to its ultimate fate after delivery.

It is easy to see why certain organs are inherently fascinating. Take, for example, the eye. How did evolution, even with millions of years at its disposal, manage to design such a complex and intricate structure? Then there is the heart, a piece of muscle that continues to beat uninterrupted for decades. Or the brain with its elusive concepts such as 'reasoning' and 'consciousness' that have kept it at the forefront of the public's imagination. How does the placenta compare with these titans of the anatomical world? In my view, very favourably. The placenta can also lay claim to many exceptional features not found in other organs of the body. These will become apparent as the book unfolds.

Many people do not have a clear idea what the placenta is, let alone what it is for. There is even some confusion as to where it comes from. Is it made by the baby, by the mother, or by both? This question is likely to elicit a wide range of opinions. I know because I have already put it to the test among friends and colleagues. Surprisingly even biologists are uncertain. The answer is that it is derived entirely from the baby but is so closely attached to the mother's womb throughout pregnancy that one can be forgiven for thinking that there must be some maternal contribution to its formation. Hands up all those who answered this question correctly!

Since the placenta goes about its duties so unobtrusively, it is easily overlooked and ignored. While she cradles her newborn infant in her arms, how many mothers would pause for a moment to reflect on the organ that has made it all possible? Very few, I would imagine, for it is not customary to present the mother with her newly delivered baby and placenta at the same time. I hope I am not stepping on too many toes when I say that obstetricians themselves are also not too concerned about the placenta in spite of the fact that many of the problems they have to deal with are the direct result of placental dysfunction. To be fair though, present knowledge about the placenta is not yet at the stage when it can be translated readily into medical practice, so clinicians cannot be blamed for just getting on with the job of treating the patient without paying too much attention to how the placenta might be involved.

Works of art do not normally generate deliberations on the placenta but there is one drawing that deserves closer inspection. Ironically, it was the absence rather than the presence of a placenta that initially provoked an interest in this picture. Leonardo da Vinci, the Renaissance master painter, is credited as the first person to depict the human baby in situ inside the womb using a specimen obtained from a cadaver. This is his renowned figure of the 'fetus *in utero*', completed around 1512, which now hangs in Her Majesty the Queen's collection at Windsor Castle (see Figure 1a).[1] Viewed through the lens of a reproductive biologist rather than an art historian, I would draw the reader's attention to several interesting features in this picture.

Famed for his critical eye for detail in all his anatomical illustrations, Leonardo curiously left out the placenta in this particular drawing. Was this an oversight or was it a deliberate omission because he did not consider the placenta sufficiently important to be featured? Leonardo's subject was estimated to be about seven months gestation age, by which time the placenta should be a prominent aspect of a pregnant uterus (see Figure 1b). Leonardo though

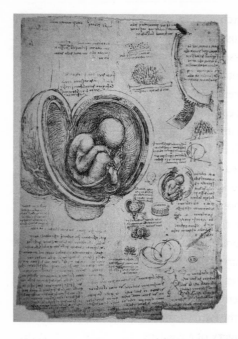

FIGURE 1 The drawing of the fetus *in utero* by Leonardo da Vinci does not include a placenta (a). (Leonardo da Vinci/ The Bridgeman Art Library/ Getty Images).

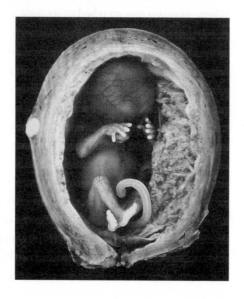

Compare this with the picture of an actual fetus *in utero* where the placenta is a conspicuous feature (b). Note also the cow's placenta and the plant seed pod in Leonardo's drawing.

4

did include a separate sketch of a placenta beside the main picture of the fetus *in utero*, but this was taken from a cow's placenta. He couldn't have made a worse choice because, as we shall see in a later chapter, the bovine placenta is one that least resembles a human placenta. Even a mouse placenta, if drawn to scale, will show a better likeness. All this, of course, was in the 16th century. Our knowledge of placental diversity has improved a great deal since then.

Leonardo's third sketch on the same page comparing the human fetus *in utero* to a plant seed in a pod was a brilliant stroke of prescience. He might not have known then how close this analogy is. Many flowering plants possess a structure called an 'endosperm' which functions just like the mammalian placenta to nourish the developing seed. Did Leonardo get the idea from the 11th century legend of *The Lamb of Tartary?* This was a tree growing in the forest of Central Asia whose seed capsule when ripe would open to reveal a little lamb within complete with a white fleece. In addition, a root sprang from the navel of the lamb, tethering it to the ground around which it grazed. This imagery of a creature, part plant, part animal, whose embryo was attached to its source of food by what appeared to be an umbilical cord, would certainly resonate with someone as curious as Leonardo. Alas, this rather wonderful mythical creation was denounced by the botanist Henry Lee in 1887 as nothing more than the early fanciful description of the cotton plant. Like all legends, the borderline between mythology and reality has become blurred with the passage of time.

The reader will notice something odd about the picture of the pregnant or gravid uterus. It shows the placenta lying alongside the baby. Since the placenta is part of the baby, it should be situated inside the baby like all its other organs. Instead, the placenta spends its entire life outside the baby but within another individual, the mother, where it is intimately attached to her womb. This peculiar association between the mother and her placenta can lead to potential conflict. The placenta, being part father like the baby, is genetically 'foreign' to the mother, making it resemble an alien transplant in her womb donated by another individual. By rights, it should be rejected by the mother's immune system like other organ transplants. And yet it is not. Why? This 'immunological paradox of pregnancy' has preoccupied immunologists for well over half a century and still the solution remains tantalizingly beyond reach.[2] If only we could find out how Nature does it so successfully during pregnancy, we might be able to use this knowledge to prolong the survival of transplanted organs. This potential clinical application is what drives research in this area besides inherent biological curiosity. It is only recently that we have started to question whether this analogy between the placenta and an organ transplant is truly justified. Yes, there are similarities, but there are also subtle differences between the two that cannot be readily reconciled. It is these differences that make the study of placental immunology so fascinating.

The mother tolerates the intrusion of the placenta remarkably well and a sort of truce is usually maintained, although not always as will be revealed in subsequent chapters. This normally peaceful co-existence between placenta and mother throughout pregnancy is one of the most extraordinary phenomena in reproductive biology. Remember that the baby itself is not attached to the uterine wall in spite of taking up much of the space inside the uterus. It lies loose within the cavity. It is the placenta that connects the baby to the mother at the site of implantation, so how the mother 'sees' her placenta rather than her baby is what determines the outcome of her pregnancy. We would do well to bear in mind this distinctive anatomical feature whenever we think about the fetal–maternal relationship. The placenta lies in between.

Nature is clever in designing the placenta to be outside the baby. In the uterus, the placenta occupies a position midway between the baby and mother, in a kind of 'no-man's land'. There is a purpose to this anatomical location. From this position the placenta is well placed to monitor and regulate all communications between mother and baby. It effectively acts as a gateway controlling what passes through. It transfers oxygen and nutrients from mother to baby and returns carbon dioxide and waste products in the reverse direction, thereby acting as the baby's lungs, kidneys, and digestive tract all at the same time throughout pregnancy. This is not all. The pregnant mother, like everyone else, is susceptible to diseases and yet, with a few notable exceptions, intrauterine infections affecting the fetus are uncommon. This is because the placenta is very efficient in screening out pathogenic organisms, but not all—some pathogens have cleverly devised strategies that can bypass the placental defence.[3] The constant sparring between host and pathogen to see who gets the upper hand is never ending. The remarkable property of the placental gateway is that it is not just an inert filter but is actively selective, encouraging the passage of substances that are beneficial to the baby while barring those that are potentially harmful.[4] It is a very sophisticated piece of machinery that has taken millennia to design. But it is not perfect. Nature, like even the best craftsman, can make mistakes. There are presently a few residual flaws in the model that allow unwanted side effects to creep in, as we shall see later. The tragic saga of Henry VIII and his wives is one such example. The placental barrier is also more porous than we thought, allowing the exchange of cells between mother and fetus. Remarkably, these cells can survive indefinitely in their new hosts.[5] All of us, therefore, are chimeras, harbouring cells within us that are not our own. The view we have of ourselves as distinct individuals may need to be modified.

Being positioned between the baby and mother has a further advantage; it allows the placenta to serve both individuals at the same time. No other organ in the body is required to perform this kind of dual function. We have already noted how the placenta serves the baby. For the mother, the placenta manufactures a variety of products that not only enable her to maintain her pregnancy, but also prepare her for events after birth, such as milk production and care of the young.[6] The human neonate is relatively helpless compared to some animal species where the offspring is already 'up and running' at birth. Postnatal maternal care is crucial for the survival of the young and is particularly well developed in primates, especially in humans. This nurturing behaviour is the result of placental hormones acting on the mother's brain.[7] Sarah Blaffer Hrdy in her book *Mothers and Others* proposes that this propensity to nurture the young was subsequently extended to the care of animals. This started Man's domestication of animals.

The relationship between the placenta and the brain is much closer than we might imagine, as the reader will see in later chapters. Several imprinted genes expressed in the placenta are also seen in the brain so that development of the two organs comes under the same genetic control.[8] Many placental hormones share structural characteristics with those made by the brain, indicating that they have evolved from an identical ancestral molecule. A further point worthy of attention is that the mother's brain can be affected by the placenta on more than one occasion. The first is when she herself is growing inside her mother's womb at the same time as when her own placenta is also developing. Later in adult life when she becomes pregnant, her brain will come under the influence of the fetal placenta she herself has conceived in her uterus. The association between placenta and brain can span two generations. In this way, the future mother already learns how to respond to placental signals even before she herself is born.

Assembling substances, such as hormones, for our own use is what normally occurs in our body. To make products for use by another person requires a much more complex organization. The latter is what the placenta has to do. For this task the placenta actually co-opts the mother to help. In this way, the baby, via its placenta, manipulates the mother's physiology and behaviour for its own benefit both during and after pregnancy. This is an extraordinary coup by the baby. To do all these things at the same time, the placenta needs to be an expert in multitasking, which it is. How does the mother feel about having her bodily functions usurped by the placenta? Is she grateful for the help or is she resentful? We shall see.

The events which allow the placenta to form outside the baby begin very early in development, from the moment the egg is fertilized by a sperm.[9] The fertilized egg immediately divides into two distinct populations of cells: embryonic and extraembryonic. The embryonic cells are destined to become the baby proper while the extraembryonic lineage will contribute towards the formation of the placenta. Thus, the fates of the two primitive cell populations are already committed. Over 80 per cent of the cells formed are extraembryonic, which goes to show that creating the placenta is even more critical than designing the embryo. This early separation allows the placenta to pursue and to organize its own programme of development totally independent of the baby. From then on, its life is not influenced by the baby. Indeed, the placenta dominates the life of the baby rather than the other way round. If the baby were to die, the placenta can live on, but not vice versa. Its relatively short existence of approximately 270 days is already pre-programmed right from the start. Its whole life is sped up relative to that of the baby. This placental autonomy holds the key to a successful pregnancy.

Our genetic inheritance comes from both parents. We would expect the genes from father and mother to contribute equally to form the placenta. They do not. In fact, they oppose each other. This is due to an unusual mechanism of gene control known as 'genomic imprinting'.[10] Paternal genes promote growth of the placenta while maternal genes restrain this growth. Without the limiting influence of maternal genes, there would be abnormal overgrowth of the placenta. The corollary is that without paternal genes there would be no placental growth at all. This has been described as parental 'conflict' or 'tug-o-war'. The father wants a big placenta to access the maximum amount of food from the mother to feed the baby, whereas the mother restricts this predatory activity in order to conserve her resources for the sake of her own health. In every pregnancy there is a struggle between father and mother to gain the upper hand. To have parents argue about who should be in charge of making the placenta seems like a petty domestic quarrel. Biologists are trying to understand the origin of this squabble. The jury is still out as to the reason why it has evolved.

Because genes from both parents are necessary for formation of the placenta, 'parthenogenesis'—development of an egg without fertilization by a sperm—is not possible in mammals, including humans. The example most often quoted as the exception to the rule is the immaculate conception by the Virgin Mary, for she herself confessed, 'how can this be for I know not man?' Furthermore, Jesus is male, and it is difficult to explain how this might have arisen from a parthenogenic conception because eggs only have the female X chromosome. Whether all this is due to divine intervention or has a biological explanation I will leave for others to decide. Noah, at least, had the foresight to recruit pairs of animals rather than singletons into his Ark. There is also the French aristocrat, Madeleine d'Auvermont, who in 1637 found herself in the unenviable position of having to explain how she could have given birth to a son in spite of her husband being away for four years. She claimed that her longing for her husband was so intense that she managed to conceive just by imagining he was with her. Her answer was accepted by the authorities and her son was declared to be legitimate. There is, of course, a more prosaic explanation than parthenogenetic conception, but let's not spoil her story. Conception by sperm alone without an egg is also a non-starter. In the story by Euripides (484 BC) Jason, in the midst of his marital problems, lamented to his wife Medea that life would be much better if men were able to conceive without the aid of women. This, of course, is biological wishful thinking, although there was a time not so very long ago when it was thought that a preformed baby, the 'homunculus' (Latin = little human) already existed in each individual sperm.

The term 'placenta' was first used by Realdus Columbus in his book *De Re Anatomica* published in 1559.[11] Until then there was no specific name for it and it was simply called the 'afterbirth'. Description of this 'afterbirth' was already well documented in ancient literature, including the Old Testament where it was referred to as the 'Seat of the Soul' or the 'Bundle of Life'. 'Placenta' is a Latin word derived from the Greek '*placous*', meaning a flat plate or flat cake. The anatomical term is 'discoid' because of its shape (see Figure 2). This is a description of the human placenta. Placentas from other animals come in a variety of shapes and sizes. This is where Leonardo made his mistake. Some animals have multiple placentas scattered all over the uterus while others have a circular placenta that spreads round the whole of the inside of the uterus. This variation is, in itself, rather intriguing and has been a source of constant fascination for comparative anatomists and evolutionary biologists.[12] Why has

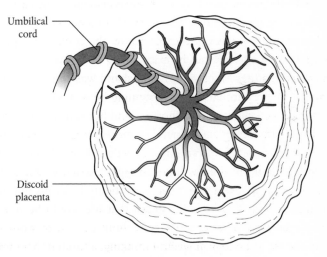

Umbilical cord

Discoid placenta

FIGURE 2 The 'flat cake' or discoid human placenta.

Nature played so many enigma variations on the theme of placental diversity? We still do not have a clear explanation for this curious finding since all other organs in the body are, more or less, the same in form and function throughout the animal kingdom. We mammals seem to breathe in air, digest food, and excrete waste products using very similar types of organs, but the way we nurture our young in the womb is vastly different. Why? What are the selective forces in evolution that might have led to each species developing its own type of placenta?

The environment must play a dominant role. Since reproductive fitness is vital for survival of the species, it is imperative that constant and rapid adaptation be made to the design of the placenta to keep pace with changes in habitat. Those species that do not adapt face extinction. As the Red Queen said to Alice in *Through the Looking Glass*, 'it takes all the running you can do to keep in the same place'. Evolution, however, does not follow any specific direction. There is a lot of trial and error strewn with blind alleys. The varied placental types we see now in present-day animals could be the result of this random progression. Characteristics which conferred a slight advantage in a particular environmental niche would be preferred and retained, while others were discarded.

Compared to other mammals, the human placenta is the most intrusive to the mother, penetrating deep into the lining of her womb and even destroying her blood vessels along the way in its quest to obtain the optimum supply of nutrients for the baby.[13] This kind of aggressive behaviour is usually only associated with cancer. Besides invasiveness, the placenta shares many other characteristics with cancer as we shall see in later chapters. This resemblance invites speculation as to the nature of cancer itself. The question is not so much why the placenta behaves like cancer but why cancer behaves like the placenta. The genes operating in the placenta during pregnancy are normally switched off in other cells of the body. Is cancer, then, an inappropriate switching on of these placental genes in other organs at the wrong time and place? If so, who has pulled the switch? Has this occurred by chance or are cancer cells doing this on purpose in order to imitate the survival tactics used by the placenta? These are all questions posed by both cancer biologists and placental biologists but coming from opposite directions.

Because of the aggressive nature of the human placenta, the mother needs to make some adjustments to accommodate it. We cannot view the placenta in isolation without including the mother's side of the story. She copes by transforming the inner lining of her pregnant uterus into a specialized layer which is able to regulate the activity of the placenta.[14] This layer has to perform two opposing functions simultaneously. It has to restrain the placenta from penetrating too deeply into the substance of the uterus, leading to serious obstetrical problems. But at the same time it must allow some degree of invasion, otherwise the placenta cannot implant properly. The balance between acceptance or rejection is a delicate one. Excessive placental invasion is bad but so is too little. Inadequate invasion will not allow the placenta to access enough nutrients from the mother to sustain its own development as well as that of the fetus. This leads to important pregnancy diseases, such as miscarriage, intrauterine growth restriction, and pre-eclampsia. Furthermore, if the baby manages to survive, the legacy of malnutrition *in utero* will linger on into adulthood when it becomes highly susceptible to a variety of diseases, a phenomenon known as 'fetal programming'. Experience in the womb can affect the whole future well-being of an individual. This offers a new insight into life before birth.

Human placentation is a highly dangerous process, inclined to myriad complications. Other species make do with a simpler and less invasive kind of placenta that is not so risky. With humans appearing so late in the evolutionary tree, it is pertinent to ask why we have not upgraded to a better placental model. Is the human placenta already so well designed that no further modification is needed or are we trapped in an out-of-date model from which there is no escape? Has development of the human placenta reached a cul-de-sac? Indeed, human reproduction is altogether not very efficient. This might be a paradoxical statement to make in view of the population explosion worldwide. It should be pointed out though, that this increase is due more to a fall in mortality than to a rise in reproductive success.

I have studied the human placenta for most of my academic career and over the years have come to look upon it with increasing awe and wonderment. The formation of the placenta is, without doubt, a significant turning point in human evolution. Without the placenta, the whole course of human history might have been different. For a start we would still be laying eggs instead of giving birth to live babies.[15] The long and tortuous evolutionary road which led from egg-laying reptiles to present-day placental mammals like ourselves was an epic journey spanning over 300 million years. Some species did not complete the journey but branched off at various time points along the way. Two such groups, the Monotremes[16] and Marsupials[17] are still with us today. These creatures are immensely interesting because they provide us with a glimpse into the past as to how we ourselves might have reproduced if we had not continued along the route towards developing a placenta. But apart from these two exceptions all present-day mammals, including humans, belong to the Eutherians, whose hallmark is the presence of a well-developed placenta. The widespread adoption of the placental mode of reproduction is compelling evidence for the advantages it must confer, otherwise evolution would not have encouraged its development. As we saw earlier, even flowering plants have decided that the development of a placenta-like endosperm to nourish the growing seed is the best way forward. For members of two such distinct kingdoms, such as plants and animals, to hit on the same idea is a classic example of convergent evolution.

Choosing the right way to reproduce that best fits with environmental conditions is the most important factor in sustaining the

continuity of a species. The phrase 'survival of the fittest' was originally attributed to Darwin's evolutionary theory but was, in fact,
coined by the social philosopher Herbert Spencer in 1864. It has
often been criticized for harbouring elitist connotations that are
somewhat distasteful. But to biologists, the term fitness refers simply
to reproductive fitness. There are no social or moral overtones.
Those who are most efficient in breeding will win the evolutionary
race. They will add to the gene pool and the species will live on.
Reproductive success takes precedence over everything, including
the fight for survival. Indeed, if breeding and survival were to come
into conflict, it is generally the latter that will be sacrificed. Take, for
example, the peacock's tail, where the attraction of a mate has priority over the unwanted attention of a predator. A beautiful and conspicuous tail is a double-edged sword. Advertising can be harmful
to health but the peacock is willing to take the chance. For him,
sexual triumph is paramount. While the male struts and preens, it is
the female who will choose because she is the one who will make
the major commitment in time and resources throughout pregnancy. Robert Trivers in 1972 formulated the theory of parental in-
vestment and, in mammals, the female invests more heavily than
the male for each offspring. In so doing, she becomes the driving
force behind the evolutionary process.

The exciting thing about placental research is that it touches on
issues beyond reproduction. The placenta shares many features with
organ transplants and also with cancer, two other major unsolved
biological mysteries of today. A common thread links their natural
history. If we succeed in understanding one, we should be able to
understand the others.

2

In the Beginning

Soon after our planet was first formed around 4.6 billion years ago, its surface was almost entirely covered with water. Even today, Earth remains a 'water' planet. The continents are dwarfed by vast tracts of ocean which is clearly evident when Earth is seen from space. And this does not include all the water hidden beneath the surface. It is not surprising then, that all forms of life started in the sea, from simple unicellular organisms like bacteria to more complex multicellular ones from which we are all descended. Preserving this aquatic legacy, the human baby continues to be cocooned in 'a bag of water', the amniotic sac, throughout its life in the womb.

Amphibians arrived on the scene during the Devonian period 410–355 million years ago. They were the first group of animals to emerge from the sea and venture on to land. Living examples of amphibians are frogs, toads, salamanders, and newts. But amphibians have not solved all the problems associated with terrestrial life. The major hurdle is reproduction. Although they can live on land, they are still restricted to life near water because they have to return to water to breed. Frogs' eggs for example, have a clear jelly-like coating which is susceptible to drying and their young larvae (tadpoles) which hatch from the eggs can only survive in water because they pass through an aquatic phase in their development when they have fish-like gills for breathing.

Following on from the amphibians came the reptiles, who had two major inventions which eliminated reliance on open water for reproduction. The first of these was the development of sexual organs by the male that could deliver sperm directly into the female's body, a process known as intromission, rather than just scattering them into the surrounding water as amphibians do. This saves a lot of wastage of sperm and increases the opportunity for a sperm to meet up with an egg. Fertilization inside the female becomes less chancy. The environment inside the female reproductive tract is also more suitable for sperm survival compared to outside. Another advantage of internal over external fertilization is that, in the latter, the coating around the egg has to be a compromise between offering protection for the egg while at the same time still permitting the entry of sperm, whereas in the former the egg coat can be added after sperm has penetrated. There is then no longer any constraint on the thickness or composition of the eggshell.

The second innovation which distinguishes reptiles from fishes and amphibians is the production of the 'amniote' or 'cleidoic' (enclosed) egg by the female. In this new type of egg, the embryo is independent of its surroundings, protected by an eggshell that can prevent desiccation but is sufficiently porous to allow gaseous exchange. Getting rid of waste products is more problematic. Fishes and amphibians excrete their waste nitrogen products as urea. This substance is highly toxic but is very soluble so will rapidly diffuse away in the large volume of water surrounding the eggs of these species. But for an embryo inside a cleidoic egg to produce urea would be disastrous because its waste product cannot escape from the closed system. The solution is to excrete nitrogenous waste as insoluble uric acid instead of soluble urea, which is then stored inside the egg until it hatches. So for evolution to be successful, two things needed to occur simultaneously. Development of the cleidoic egg can only happen when there is also a change in the mechanism for the excretion of waste products by the embryo. The uric acid waste is stored in the 'allantois'—a sac-like pouch that has sprouted from the rear end of the gut during development of the embryo. The allantois also has a plentiful supply of blood vessels lying just beneath the eggshell that will act as conduits for gaseous exchange between the embryo and its outside world. The allantois has become the embryo's 'dustbin' as well as its 'lung'.

As a result, an independent viable structure was formed which permitted the embryo to feed, breath, and excrete safely inside the egg. These adaptations emancipated reptiles from the necessity to breed in water and allowed them to colonize land, sea, and air, making them the most successful species during the Mesozoic era 250–65 million years ago, which has become known as the 'Age of Reptiles'. This lasted until something occurred which eliminated many of them, particularly the larger ones (e.g. dinosaurs), at the end of the Cretaceous and the beginning of the Tertiary period 65 million years ago. This is known as the Cretaceous/Tertiary (KT) extinction. Earth scientists are still arguing as to what might have happened.

Although not yet fully accepted, the most popular theory at present is that it was triggered by a huge asteroid colliding with Earth, gouging out a crater 60 miles across and 18 miles deep which is still visible today in Chicxulub, Mexico. The asteroid hit the Earth at such an astonishing speed that the energy released on impact was estimated to be equivalent to more than a billion times that of a nuclear bomb. Although the collision caused catastrophic devastation, such as huge fires and severe earthquakes, it was the aftermath of the impact which was ultimately responsible for the demise of many species. Vast clouds of debris from the blast were thrown up high into the atmosphere, blocking out the sun's rays and blanketing the Earth in decades of darkness. Without the sun, plants were unable to photosynthesize and gradually died out. This was what ended the reign of the dinosaurs and paved the way for the eventual ascent of mammals to become the dominant species on the planet. Sad for the dinosaurs but fortunate for us mammals. Within a mere two to three million years, mammals had radiated explosively throughout the world, some of which flourished while others died out along the way. For the first time in their evolutionary history, mammals of middle to large size appeared on the scene. Before that, only small mammals, most of which were no larger than modern mice, existed. This was because of environmental competition with the large dinosaurs. Following the demise of the gigantic reptiles, mammals spread out so successfully that the present geological era (Cenozoic) is now known as the 'Age of Mammals'.

Would mammals have emerged if dinosaurs had not been eliminated? Some palaeontologists believe the answer is 'yes'. Catastrophic events merely hastened the process. It might be worthwhile mentioning that there are thousands of asteroids presently orbiting nearby. Sooner or later, one of these will collide with Earth. What happens then? Who will be left?

Most reptiles laid eggs and many still do. The first step on the road from oviparity (egg laying) to viviparity (birth of live babies) was taken by some reptiles who retained the fertilized egg inside the female where it eventually hatched in the womb or uterus. In the most primitive versions of viviparity, the embryo was nourished and underwent all its development within the amniote egg until it hatched inside the mother's womb. This is seen in many reptiles today (e.g. the adder). Because this form of reproduction falls halfway between oviparity and viviparity, it is termed 'ovo-viviparity'. A more advanced form of viviparity followed, in which the fertilized egg made contact and attached itself to the wall of the uterus. The eggshell became thinner and this area of contact was subsequently modified into a specialized organ, the placenta. This was the beginning of the placental form of reproduction. The establishment of the placenta made it possible for the embryo to now access nutrients directly from the mother. There are different ways of doing this, which accounts for the wide variety of placental types we see today. Occasionally, there is a more gruesome way by which embryos obtain nutrients inside the womb, not from the mother but at the expense of other embryos. In some species of shark one or two dominant embryos will devour their weaker siblings—a kind of intrauterine cannibalism.

The principal advantage of viviparity is that the young can be nurtured for long periods in the safety of the womb away from predators and other hazards of the exterior environment, but the mother has to bear the physical burden of carrying the young inside her until they are born. This may explain why birds have continued to lay eggs, because an oviparous mode of reproduction is obviously more suited to their life in the air.

The discovery in 2008/2009 of the fossilized remains of two pregnant primitive fishes has revised backwards the estimated time for the origin of viviparity. *Materpiscis attenboroughi* (in honour of Sir David Attenborough) and *Incisoscutum ritchieri* were each found with a fetus attached to the mother fish by an umbilical cord. These two fishes belonged to the family of Placoderms which were ancestors of modern sharks and bony fishes of today. They date back to the Devonian period 380 million years ago. Some male Placoderms had appendages that could clasp the female tightly during mating, suggesting they practised intromission with delivery of sperm directly into the female. It seems that viviparity had evolved much earlier than previously thought, even before the emergence of reptiles. For such ancient vertebrates to adopt the viviparous method of reproduction emphasizes the advantage it must confer.

The Mesozoic were exciting times, an era that ushered in the dawn of mammals. Some 200–250 million years ago, a group of reptiles known as Therapsids split off from the rest by introducing three further innovations which were to make a dramatic impact on the next phase of evolution. The first of these was the ability to generate and regulate their own internal body temperature, a process known as homeothermy. This was a significant improvement on amphibians and other reptiles that still depended on the sun's rays to warm up their bodies before they could be active. This is the main difference between 'warm-blooded' and 'cold-blooded' animals. Following on from homeothermy was the development of a covering of body hair to provide insulation so that the heat generated inside the body was not lost to the exterior. The third innovation was the production of breast milk via mammary glands: lactation. The term mammal is derived from the Latin 'mamma', meaning breast. These Therapsid reptiles were the ancestors of all modern mammals. The production of milk is not only a way of providing food for the offspring but the associated parental care involved in suckling the young and the bond created between mother and child is a typical characteristic of mammals. This latter feature is especially well developed in primates, such as humans. Lactation, together with homeothermy and viviparity, is responsible for the major breakthrough in mammalian evolution.

During the course of evolution, many species gradually moved from living in water to making a home on land. Curiously, some terrestrial mammals, particularly the larger ones like the whale, dolphin, seal, and hippopotamus, decided to migrate in the reverse direction from land back to water. All these mammals have adapted to an aquatic existence by losing their hair covering but have developed a compensatory layer of fat underneath their skin (blubber) for insulation. Hair or fur is not an efficient insulator in water. Whales and dolphins have been aquatic for over 70 million years and have never returned to land. They even breed in water. To give birth and to suckle an air-breathing offspring under water is an impressive adaptation. Milk in whales has a very high fat content which contributes to the formation of blubber in the baby. Seals, on the other hand, still need to come onto land to breed as they only became aquatic relatively late, around 25–30 million years ago, and have not quite made the complete transition to water like whales and dolphins. Seals, whales, and dolphins have modified their limbs into fins but hippos still have limbs. With their semi-aquatic lifestyle and ungainly gait on land, hippos can be considered as having accomplished half their evolutionary journey from land back to water. Whales today continue to develop hind limb buds during embryonic development but this stops around the fifth week of gestation when the limb buds are reabsorbed. This brief glimpse into their evolutionary past confirms that the ancestors of whales were originally terrestrial animals. While whales have discarded their terrestrial characteristics like hair and limbs, they have retained their mammalian traits such as lactation and suckling.

Another group of large mammals that has taken to the sea are the sea-cows and dugongs, which are related to elephants. Sightings of these creatures appear to be the origin of the myths of mermaids described by sailors. But any resemblance to fair maidens would be hard to imagine since these are large, unattractive creatures weighing up to a ton. Maybe long periods at sea without female company clouded the sailors' aesthetic judgement. The fact that reverse migration from land back to water has not occurred among smaller mammals such as rodents, suggests that the better buoyancy offered by water could be the inducement to return to an aquatic life. One only has to see how clumsy seals and hippos are on land compared to the ease with which they cleave through water.

On the subject of buoyancy, it is curious that human babies also have a thick layer of subcutaneous fat, much more than in other primates. This is reflected in the cherubic appearance of human babies contrasted with the rather wrinkled expression of newborn monkeys.

With this layer of fat, human babies are also highly buoyant, frequently being able to swim even before they can toddle. They have little hair covering. All these are features found in aquatic mammals. Elaine Morgan, in her book *The Aquatic Ape Hypothesis*, argued strongly that our ancestors walked out of the sea instead of descending from trees like other primates. But there might be a simpler explanation. Human babies need to accumulate a store of fat for brain growth. Fat makes up to 50 per cent of the dry weight of the brain. In humans, development of the brain does not end at birth but continues for some time postnatally and fat is needed for this phase. The human placenta is very efficient in transporting fat across from the mother to her fetus, especially during the later stage of pregnancy when up to 7 g per day are transferred. This is when the baby builds up its own fat deposits in readiness for the time when its mother's fat is no longer available. No wonder humans have the fattest babies of all primates.

Some mammals have taken to the air. These are bats. It may come as a surprise to learn that bats have a placenta very similar in design to our own, they suckle their young and female bats even menstruate. Mammals, then, have adapted to ecological niches on land, sea, and air, replacing the dominance reptiles once had.

The early mammals which evolved from the Therapsid reptiles continued to diverge from each other and ultimately became three distinct groups: **monotremes**, **marsupials**, and **eutherians** which differ from each other in the way they reproduce (see Figure 3). Monotremes, so called because they have only one common excretory opening, the cloaca, for both the urogenital and digestive systems, split off from the marsupials and eutherians 166 million years ago, but surprisingly continue to share over 80 per cent of their genes with the other two groups. The crucial difference is that monotremes continue to lay eggs. They are the only mammals to do so. Current examples are the duck-billed platypus and the echidna or spiny anteater, both considered to be the most primitive of

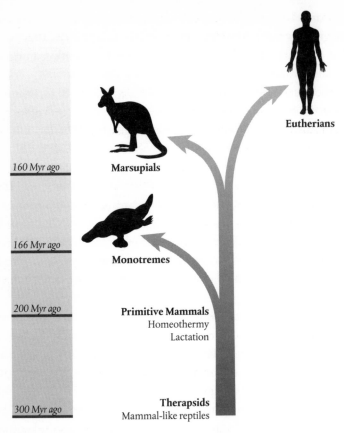

160 Myr ago

Marsupials

166 Myr ago

Monotremes

200 Myr ago

Primitive Mammals
Homeothermy
Lactation

300 Myr ago

Therapsids
Mammal-like reptiles

Eutherians

FIGURE 3 Evolution of monotremes, marsupials, and eutherians.

mammals. Only three living species exist, all in Australasia. Not all mammals, then, are viviparous. However, unlike oviparous reptiles and birds, the early development of the monotreme embryo within the egg takes place over several weeks inside the uterus as the eggshell stretches to accommodate the growing embryo. During this period, the embryo is nourished by secretions provided by the uterus which diffuse across the porous eggshell. So, although not viviparous, the mother makes a significant contribution to feeding the growing embryo inside the uterus even in oviparous monotremes. The monotreme mode of reproduction is halfway between oviparity and viviparity.

The duck-billed platypus[18] (Greek: *platypous* = flat feet) (see Figure 4) created quite a stir when the first specimens were sent to Europe from Australia at the end of the 18th century because of its bizarre appearance. It's a tease. 'Neither fish nor fowl' was how it was described. It seemed to be made up of odd bits from different animals stuck together at random. There was even an idea that it could be the rare successful product of inter-breeding between different species of animals. Indeed, scientists were so sceptical when they first saw it that they actually cut up one specimen to look for incriminating stitches where the various parts could have been joined in case it was a hoax! The platypus also cannot decide what kind of animal it wants to be biologically. It mixes its mammalian characteristics with others it has retained from its reptilian ancestors. Like other mammals, it possesses a coat of fur and the ability to generate its own internal body temperature. It produces milk and suckles its young, although its mammary apparatus is rather primitive: the mother has no defined nipples and the young just lap up the milk which oozes through a patch of skin. But it continues to lay eggs like reptiles and has even retained the ability to secrete venom, another

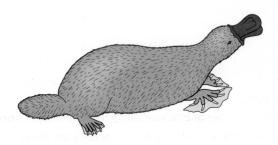

FIGURE 4 The duck-billed platypus (*Ornithorhynchus anatinus*).

legacy from its reptilian past, which the male can deliver from spurs on its hind legs. The genes that produce this venom are very similar to those found in snakes. No other mammals are known to secrete snake-like venom.

From the time the platypus was first identified, to establishing the fact that it was oviparous, took an astonishing 90 years. It is a fascinating story, beautifully captured in the book *Platypus* by Ann Moyal. The scientific hub at that time was in Europe, while the subject of research was an inhabitant from the other side of the world. Communication was not easy. It was a young Cambridge embryologist, William Caldwell, who solved the riddle in 1884 when he found one egg already laid and a second egg at the mouth of the uterus in a female platypus waiting to be laid. He sent a telegram to the British Association for the Advancement of Science that was meeting in Canada at the time. The announcement of his findings stunned the scientific world. 'Monotremes oviparous, ovum meroblastic' was the message. 'Meroblastic' describes the incomplete cell division used by eggs with a lot of yolk, such as those laid by reptiles and birds. Eggs from placental mammals with little yolk divide by a complete 'holoblastic' mechanism. Thus, Caldwell's findings have a double significance. Not only do monotremes lay eggs but their eggs resemble those produced by reptiles and birds rather than those of placental mammals. So the question that had puzzled scientists for so long was finally answered, but not without a price. Thousands of platypuses were slaughtered in the quest for evidence. Ironically, the Australian Aborigines probably knew the answer all along but their word was judged to be unreliable. For some reason Aborigines do not eat platypus, so these animals are never hunted.

Caldwell was a graduate student working in Cambridge University's Department of Physiology at what is known as the Downing Site. At that time, the department was already well established at the forefront of embryological research. This strong tradition continues today. The 2010 Nobel Laureate, Bob Edwards, who pioneered *in vitro* fertilization, was a member of the department when he did his ground-breaking work. Azim Surani, the present holder of the Marshall Chair of Reproduction, is based in this department. He is the scientist who created mouse embryos containing either only paternal or maternal genes which led to the theory of 'genomic imprinting' (see Chapter 4). The recently established Centre for Trophoblast Research is also housed in this department. This is the only centre in the world that is devoted entirely to the study of the placenta (see Chapter 3). Caldwell's trip to Australia was partly funded by the Balfour Fund which was established to commemorate Professor Francis Maitland Balfour, the then Head of Department, who was killed in a tragic mountaineering accident at the young age of 31. The Fund still exists. Generations of natural science and medical students have passed through the gates of the Downing Site which houses all the biological science departments. The thought that Caldwell would have walked through the same gates (or would he have cycled?) like so many of us have done and continue to do, has brought me personally much closer to the story of the platypus.

There was still one more question about the platypus that was not yet settled. The prevailing view at that time was that oviparous animals were not supposed to lactate. Only viviparous species suckled their young. There was, therefore, much debate about the glands found under the skin of the platypus that did not terminate in a nipple. Were they true mammary glands or were they just glands for lubricating the skin? Were their secretions really milk? How did the young platypus with its cumbersome duck bill suckle? These questions were eventually answered. Yes, they were mammary glands that secreted milk and the infant platypus' mouth was perfectly capable of lapping up the milk expressed onto the skin. In one go the platypus had succeeded in demolishing two firmly held beliefs of that era: that mammals did not lay eggs and that oviparity and lactation were mutually exclusive. The platypus had unequivocally set the record straight.

The female usually lays two eggs and incubates them in her nesting burrow for 10–11 days until they hatch. She then suckles her young until they leave the burrow 3–4 months later already covered with fur and able to swim. This close postnatal care is a typical characteristic of mammals so the platypus has at least acquired this mammalian trait. The newly hatched platypus is tiny, weighing less than half a gram. They need to be fed with milk. The same is true for the other monotreme, the echidna, whose young are also minute and have to be nursed for a substantial period in a 'brood pouch' near to a pair of mammary glands. Lactation, then, preceded viviparity in the evolution of mammals and was already adopted by oviparous monotremes to feed their young. This method of feeding is retained by the viviparous marsupials and eutherians. Not only is milk highly nutritious, the act of suckling creates a bond between mother and infant that enhances maternal care and nurturing, like an invisible umbilical cord continuing to bind the two together after birth. In eutherians, with a longer gestation inside the womb, the placenta takes over much of the responsibility for sustaining the fetus and breastfeeding is postponed until the postnatal period. Distinct nipples only became apparent in later viviparous species. At first, there were multiple nipples on each side of the mother's body arranged along a line drawn from armpit to groin. Some animals today (bitches) still continue with this arrangement. Others have just retained a few nipples in the abdomen (cattle) or in the chest (humans). In the rare cases of humans with extra nipples, these are always found somewhere along the two ancient lines. Even men have breasts and nipples although these are no longer needed. They are remnants of an evolutionary past. It might seem paradoxical that the initial development which led to viviparity was something happening outside the uterus like lactation, rather than inside such as the advent of placentation.

The next group of mammals, the marsupials, is equally fascinating.[19] Unlike the monotremes, which are represented by only three species, there are approximately 260 living species of marsupials, the majority of which dominate the mammalian fauna of Australasia, such as the kangaroo and opossum. A few species are also found in the Americas. Although no longer egg-laying, marsupials nevertheless have a most unusual mode of reproduction. They are viviparous and do have a placenta but the young are born very immature, naked, and blind (known as joeys) and need to be nurtured for a further period in a pouch where they obtain nutrients via breast milk secreted by the mother, hence, 'marsupium' meaning pouch. A kangaroo's baby when first born is no larger than a human baby's finger. Because the young are so minute and undeveloped, it was once thought that the pouch had replaced the womb and was the place of conception. This led to the idea of the 'external uterus', a concept which lasted for almost a century.

Can we humans replicate the kangaroo's ingenious way of caring for its very immature young in dealing with our own premature babies? Prematurity, defined as any baby born before 37 weeks of gestation, is a major cause of infant mortality worldwide. In the developed world, with access to modern technology, many of these babies can be saved. But in poorer countries, it is estimated that 90 per cent of premature babies do not survive. The World Health Organization has designed an inexpensive scheme, appropriately called the 'Kangaroo Mother Technique', whereby the premature baby is strapped skin-to-skin near to the mother's breast for 24 hours a day for warmth and ease of suckling, in an attempt to emulate the successful marsupial pouch strategy. Results are encouraging.

Marsupials have traded a long gestation period in the womb for lactation in the pouch. This is an excellent reproductive strategy for coping with the harsh and unpredictable environment of the Australian bush. By transferring the young to the pouch very early on, the vacated womb is now ready to accommodate a further pregnancy. The mother lays another egg, which becomes fertilized, but now a curious thing happens. The development of the fertilized egg in the womb is arrested because the mother is still suckling the joey in the pouch. This arrested development is known as 'diapause'. When the joey reaches a certain stage of development, it stops suckling and leaves the pouch to fend for itself in the world outside, returning only occasionally to feed from the mother's breast. The cessation of suckling acts as a trigger to restart the arrested development of the fertilized egg, waiting in the womb, into a new embryo. When born, this new joey will crawl to the pouch to suckle. The mother produces another fertilized egg, whose development is again arrested in diapause. The cycle continues. With this ingenious arrangement, the marsupial mother can nourish three youngsters simultaneously, all in different stages of development: an older, semi-independent youngster returning to feed in the pouch occasionally; a young infant joey suckling in the pouch; and a fertilized egg lying dormant in the womb waiting to develop further. If environmental conditions remain harsh, the mother jettisons the first and second of these but keeps the third option in reserve in her womb until good times come round again. Delayed implantation or diapause is not restricted to marsupials. It has evolved independently in other mammals such as bears, roe deer, bats, armadillos, and many carnivores. It must be a highly successful reproductive device for it to be adopted across so many unrelated species, although marsupials have refined it to a fine art. Humans have lost this rather useful trick.

Australia has often been referred to as 'the Lucky Country'. This is certainly true with respect to its flora and fauna. The continent drifted away from the rest of the southern land mass of Gondwana over 200 million years ago and has remained isolated ever since. Shielded from the prying eyes of migrating predators, a variety of exotic creatures flourished, such as the monotremes and marsupials. These animals have not reached an evolutionary dead end but are constantly becoming more specialized in their secluded environment. They might appear to exhibit traits reflecting our past heritage but they are not primitive animals.

The separation of marsupials from eutherians occurred around 160 million years ago. There are presently two views on the evolutionary relationship between the marsupial and eutherian mode of reproduction. One view is that marsupial reproduction evolved first, being ancestral to the eutherians, and is more primitive. The failure to find a solution to deal with the hazards of intrauterine life, such as immunological rejection by the mother (see Chapter 6), curtailed the time the fetus could spend inside the womb and it was relegated to the pouch for further development. The second view is that marsupial reproduction evolved independently. It is equally efficient compared to eutherian reproduction. Marsupial reproductive strategy is certainly well adapted to the unpredictable environment of Australia as already discussed. Marsupials have flourished in this continent because of the absence of competition for food from other large herbivorous eutherians which are so abundant elsewhere in the world.

This might be an opportune moment to revisit Darwin's theory of evolution by natural selection. He likened it to a branching tree where the stem of the tree represented the common ancestor while the branches radiating outwards were the different species derived from the main stem. The characteristics of each species had been selected by Nature because they fit in best to an environmental niche. One species is not 'better' or 'worse' than another. Evolution is not like a ladder where those from a lower rung strive to climb towards the top in a continual quest for improvement. Before Darwin, many biologists believed that evolution of life on Earth followed a master plan with each step progressing relentlessly towards a predetermined goal, which most considered to be the human species. Hence, the term 'evolution' itself is derived from the Latin 'evolutio', meaning the unravelling of a scroll. Darwin argued that this is not what evolution is all about. There is no master plan; there is no predetermined goal; there is no utopian destination. Instead, evolution follows a random course. Nature selects those species best adapted to a particular environment and these will survive and prosper. Those that do not adapt will die off along the way. Viewed from this Darwinian perspective, marsupials are not more primitive or inferior to eutherians in their reproductive strategy. They just come from a different branch of the tree. Marilyn Renfree, the Australian biologist, prefers to regard marsupials as 'alternative' mammals rather than 'lower' mammals.[20]

It is among the eutherians that the most developed placentas are found. Apart from monotremes and marsupials, all other mammals we know today including humans belong to this group. They are by far the largest of the three groups, being represented by over 4,500 species spread throughout the world. If the number of living examples reflects evolutionary success, then having well-formed placentas appears to be a distinct advantage.

The eutherian placenta comes in an almost infinite variety of shapes, sizes, and structures. Biologists have long tried to make sense of this chaotic arrangement. Why did Nature make so many different kinds of placentas? No other organ in the body shows this degree of variation. The best way to cut through this structural maze is just to view the different placentas as belonging to three basic types according to how deeply it invades the uterus (see Figure 5).[21] For example, in ruminants (cattle, sheep), the placenta merely attaches itself superficially to the surface lining of the uterus without eroding it any further. This type of placenta is known as '**epitheliochorial**' because it is in contact with the epithelial lining of the uterus. At the other extreme (rodents, humans), the placenta penetrates deep into the wall of the uterus, destroying the underlying blood vessels along the way and coming into direct contact with the mother's blood. This type of placenta is therefore called '**haemochorial**' (*haem* = blood). Epitheliochorial and haemochorial placentation represent the two extreme ends of a spectrum. There is a third type of placenta which falls in between these two categories. It is used by carnivores such as the cat

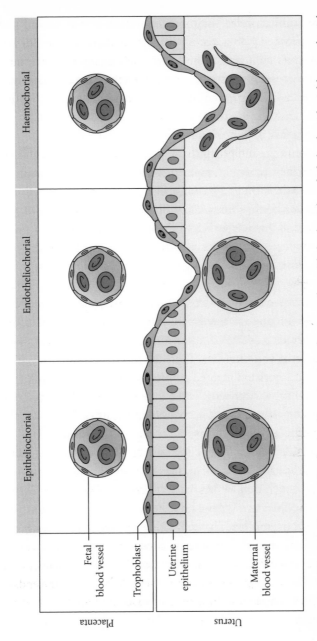

FIGURE 5 Diagram showing the depth of invasion into the uterus by epitheliochorial, endotheliochorial, and haemochorial placentas.

and dog families. Here the placenta succeeds in breeching the uterine epithelial lining and invades the underlying tissues but stops short when it gets to the uterine blood vessels. There is no further progression or destruction of these vessels as happens in the haemochorial variety. The uterine vascular network remains intact and the placenta is not bathed in maternal blood. This type of placentation is termed '**endotheliochorial**' because the placenta reaches the endothelial lining of the uterine blood vessels but no further. From the point of view of increasing degrees of invasiveness, the progression can be depicted as epitheliochorial → endotheliochorial → haemochorial. No matter what type of placenta, the one characteristic that is shared by all of them is separation of the fetal and maternal circulations. For reasons that will become clear later in Chapter 5, the two blood streams never mix.

The German anatomist, Grosser, in 1909 was the first to appreciate that the microscopic feature which distinguishes these three different types of placentas is the number of layers separating fetal from maternal bloodstreams. There are more intervening layers in the epitheliochorial compared to the haemochorial placenta with the endothelial variety somewhere in between. This is clear from Figure 5. Logically, it might be expected that the multiple layers of the epitheliochorial placenta would present a greater obstacle to the transfer of material from one circulation to the other and it is therefore less efficient. This view persisted for some time but we now know this conclusion to be incorrect. The main functional difference between the two types of placentas is the source of food they access from the mother. The epitheliochorial placenta feeds on nutrients mainly derived from fluid secreted by the uterine glands, also known as 'uterine milk' so there is no need to seek out and destroy maternal blood vessels, whereas for the haemochorial placenta, all nutrients are obtained from maternal blood. One source is as good as the other.

Haemochorial placentas are found in the most unexpected places. Unrelated species such as rodents (e.g. mice, rats, hamsters, guinea pigs) and even the hedgehog, the armadillo, and some species of bats share a similar type of placenta. The corollary to this is that not all primates possess haemochorial placentas. The more primitive primates, such as lemurs and lorises, utilize the non-invasive epitheliochorial placenta. There is no clear-cut phylogenetic pattern; that is, the type of placenta does not follow the evolutionary relationship of different species. A further puzzling feature is that some species cannot decide which variety of placenta they prefer. Some bats, for example, start off with an endotheliochorial placenta and then progress to a haemochorial one at the later stages of pregnancy. This change may be necessary to cope with the increasing demands of a fast-growing fetus.

The horse is another example whose placental type is difficult to categorize. The equine placenta is essentially non-invasive epitheliochorial and in this regard it is similar to other species with this kind of placenta, such as the pig. But in the horse, by day 36 of pregnancy, several areas of the placenta begin to penetrate the uterine epithelium into the underlying tissues to form distinctive structures known as 'endometrial cups'. The cups survive for about two months but around days 100–120 (the mare's gestation is 340 days) they start to regress and mysteriously disappear. It is not clear whether the cups are ultimately rejected by the pregnant mare or whether they simply die off spontaneously at the end of their natural lifespan. Endometrial cups are unique to the equine placenta and are not seen in any other species. The cups secrete the placental hormone equine chorionic gonadotrophin (ECG) which is the functional equivalent of human chorionic gonadotrophin (HCG) (see Chapter 7). Both hormones are for maternal use so the transient appearance of the cups might be to facilitate dissemination of ECG into the pregnant mare's circulation. The equine placenta, then, is highly unusual in being essentially non-invasive epitheliochorial but exhibiting localized areas of invasion at a specific time during pregnancy.

Sheep and goats have a different approach. Although there are no overt signs of invasion by their epitheliochorial placentas, closer examination reveals areas where placental cells have actually fused with cells of the uterine epithelium, forming hybrid cells. These hybrid cells produce placental hormones which readily diffuse into the maternal circulation. Here we have a situation where placental cells sneakily commandeer the production machinery within maternal uterine cells to synthesize the placenta's own products for delivery back to the mother. This must be a supreme example of stealth in placental strategy. We usually think of the placenta as an organ which enables the fetus to access nutrients from the mother. The examples we have just seen remind us that the placenta has also been designed to facilitate delivery of products in the reverse direction from fetus to mother. Don't forget, the placenta needs to serve two masters.

Because the laboratory mouse placenta is haemochorial, this species is the most frequently used experimental animal for placenta research. But mice are not ideal. Although haemochorial, the pattern and depth of invasion into the uterus by the mouse placenta is not identical to that of humans. Also, gestation in the mouse is only 21 days compared to the nine months of human pregnancy so the functional requirements of the two placentas are unlikely to be the same. The human mother has to shoulder the burden of pregnancy much longer than the mouse. What's more, mice normally give birth to a litter of offspring while multiple births in humans are the exception rather than the rule. Attempts to extrapolate data obtained from experiments with mice to humans have led to a great deal of confusion and misinformation. Even the placentas of Old and New World monkeys are not the same as humans and it is not until we get to our nearest cousins, the Great Apes, that some similarity is seen. There is not a ready supply of apes for research and permission to use them is nigh on impossible to obtain.

Alexander Pope, in his 18th-century poem, asserts that 'the proper study of Mankind is Man'. Fine in theory but difficult to achieve in practice. Human studies can be done either directly in the patient (*in vivo*) or by using human placental tissue cultured in the laboratory (*in vitro*). The former is ethically dubious, while the latter is fraught with technical problems (see Chapter 3). This is why progress tends to be rather slow. That is not to say that observations on animals have no value at all.[22] For example, genes identified as being involved in the development of the mouse placenta can point the way in the search for similar genes in humans because they tend to be conserved across species. The 'knock out' (KO) mice are particularly useful. These are mice whose genes are artificially inactivated to see what effects this procedure has on the animal. As we shall see in later chapters, disruption of selected genes responsible for placental development can reveal what their normal functions are. In this way, we can slowly build up a catalogue of important placental genes and what they do. Obviously, this kind of experiment cannot be carried out in humans. So research on the human placenta should continue in tandem with those on other species. Such studies will not only provide us with a broader view of what is happening in the rest of the animal world but will also give us a better understanding of our own.

Among all the species with haemochorial placentas, the human placenta is the most invasive of all. The advantage is obvious: it establishes the most intimate contact between the developing embryo and its source of food from the mother. The disadvantage, of course, is that it is very dangerous to the mother because of the potential for haemorrhage and perforation of the uterus. Because the haemochorial placenta is so costly to the mother, it is a general rule that animals with this type of placenta tend to produce small, helpless or altricial offspring after a short gestation: spend the minimum time in the womb and get the baby out quickly. Mice are good examples. In contrast, the epitheliochorial placenta is less demanding, allowing the mother to sustain its presence longer *in utero* with the birth usually of a single, large, precocial neonate as seen in hoofed animals.

The one exception seems to be our own species. Humans have the most invasive type of placenta and also the longest gestation period. Why have humans evolved this bizarre and almost surreal combination? What are the selective pressures during evolution that made us adopt this somewhat risky model? The most often quoted explanation is that this kind of placenta is required to ensure adequate nutrition and oxygen supply for the development of the large brain of the human embryo. It is calculated that, during pregnancy, 60 per cent of total nutrients are directed to the development of the human fetal brain compared to around 20 per cent in other mammalian species. But large brain size is not always associated with haemochorial placentas. Dolphins, especially the 'bottlenose' variety, the mammalian species with the second largest brain after humans, make do with the non-invasive epitheliochorial placenta. Not only do dolphins have large brains, the anatomical structures associated with intelligence are very similar to those of humans. Indeed, behavioural studies have established dolphins to be the second brightest creatures after humans, deposing chimpanzees who were once accorded the runner-up prize in the intelligence stakes. Perhaps brain size relative to body size is a better measure than brain size per se. Looked at from this angle, the mouse brain is the largest, making up 10 per cent of body mass. Next come humans with 2 per cent and dolphins with 1 per cent. Chimpanzees are low down on the list, with the brain comprising only 0.8 per cent of its body mass. And yet, chimpanzees have a deeply invasive type of haemochorial placentation very similar to humans. Why do chimpanzees need to access so many nutrients for development of such a modest brain? So the theory that the human placental model has evolved to sustain the development of a large brain is not entirely supported by observations in the dolphin and the chimpanzee. At present, there are no plausible alternative explanations. Until we find one, this existing theory has to suffice.

The fact that haemochorial placentas are found in such unrelated species as mice, hedgehogs, armadillos, bats, and humans suggests there must have been strong selective pressures during evolution that favoured this condition. This is an example of 'convergent' evolution in which unrelated animal species facing the same problem develop similar solutions. But what is the problem that is shared by all these species? We must not assume that the haemochorial placenta is necessarily the most advanced in evolutionary terms. We do not know which type of placenta evolved first.[23] Intuitively, we would suppose that the superficial attachment of an epitheliochorial placenta is likely to be the more primitive, followed by an increasing degree of invasion as seen in the endotheliochorial and then the haemochorial placenta. In our arrogance and from our anthropocentric perspective, we would like to believe that the human deeply invasive placenta must be the most sophisticated and the most advanced of them all. But the human placenta might not be the finest of them all.[24]

Current evidence points to either the endotheliochorial or the haemochorial placenta as the ancestral form and the epitheliochorial type being a later, more modern, derivative. This is unexpected. Placentas are made up entirely of soft tissue, so no fossil remains are available. The past has to be inferred, such as by molecular phylogenetic studies.[25] This is a science that tries to assign today's animals into different groups, known as 'clades', according to their evolutionary relationships. While invasive endotheliochorial and haemochorial types of placentation are seen in all clades, epitheliochorial placentas are present only in the more modern ones, indicating that this type of placenta evolved later.[26] *Eomaia* was a tiny furry animal that lived 124 million years ago. Its skeletal structure and teeth pattern point to it being an insectivore that lived in trees. Present day insectivores of a similar type all have an invasive endotheliochorial or haemochorial placenta. It looks as if our ancestral placenta started off being invasive and then changed its mind and withdrew to become less invasive. We do not know why this happened. Environmental pressures from different habitats could provide the evolutionary stimulus to select the most appropriate placental type for individual species. Since the discovery of *Eomaia*, the fossils of another, even more ancient, eutherian has now been found in China.[27] Named *Juramaia*, it is estimated to be 160 million years old, and therefore predates *Eomaia* by 36 million years. Since *Juramaia* is now the oldest recorded eutherian, the split between marsupials such as kangaroos and eutherians like ourselves must have occurred at least 160 million years ago during the Jurassic period, much further back in evolutionary time than first thought.

The debate about which is the ancestral placental model is not a new one. It has been rumbling on since the end of the 19th and beginning of the 20th century. For example, Hubrecht working on the haemochorial placenta of the hedgehog considered this to be the most primitive, whereas Mossman favoured the endothelial variety in carnivores as the ancestral form. Today, the jury is still out as to which one of these came first, but most people are agreed that the non-invasive epitheliochorial variety was probably the last to appear on the scene. Certainly, the epitheliochorial placenta is remarkably efficient, as can be seen in the successful reproduction of ungulates (hoofed animals such as sheep, cows, goats, horses) with their precocial offspring which are born relatively mature and already able to run and fend for themselves compared to the altricial offspring of many species with haemochorial placentas such as mice and humans which are incapable of independent existence when born. There is the story of the physiologist, Sir John Hammond, who, together with numerous assistants, spent a considerable time attempting to catch a newborn foal. Afterwards, an exhausted assistant remarked to another, 'not a bad advertisement for a multiple layered placenta!'

The epitheliochorial placenta does not limit fetal growth and development. The assumption that the deeper the invasion the easier it is to access food from the mother may not be correct. The surface area available for nutrient and gaseous exchange is probably more important. A diffuse epitheliochorial placenta spread over a wide area inside the uterus can be better at absorption than a deeply invasive but localized haemochorial one. Efficiency of nutrient transport, therefore, cannot be the sole explanation for the divergence of placental types. Other factors are likely to come into play. An important one to consider is the mother's immune response to the placenta (see Chapter 6). The deeper the invasion, the more chance of triggering a potentially damaging reaction. Establishing just a superficial contact between placenta and uterus, as in the epitheliochorial placenta, could be a very clever move. At the end of the day, all the debate over which placental type evolved first might not be too meaningful. Unlike present day technological products, the latest placental model is not necessarily the best. Evolution does not progress in a straight line. One type of placenta is more suitable than another in its own environmental setting.

The evolutionary history of ungulates and humans has followed different pathways. The former have evolved to live in the wide open spaces of a grassland habitat where there is little shelter from predators. The ability to run fast is, therefore, a big advantage. Even their young need to be born 'up and running' because they need to keep up with the herd. Their sight and hearing are also well developed in order to recognize the mother. This is important as ungulate mothers only care for their own offspring and not for others. Indeed, unrelated young who try to suckle will be peremptorily rejected. The reproductive strategy in humans is not so straightforward. There are conflicting evolutionary pressures. Compared to apes, the head of the human baby in the womb is large due to the increasing size of the human brain. At the same time, the adaptation by humans around 6 million years ago of an upright posture and walking on two legs has narrowed the pelvis (see Figure 6). The combination of the two is an awkward marriage. It has made the birth process in humans a finely balanced affair. The birth canal now is almost the same width as the baby's head. With natural variations in pelvic size, some women are bound to have one that is too small for the job. There is no room for error. Protracted and very powerful contractions are required to push the baby out. These contractions can lead to intermittent interruption of blood flow to the baby in the womb which will cause fetal distress. A change in fetal heart rate is the first sign of this distress which is why it is important to monitor the fetal heart beat constantly during a difficult labour. Obstructed delivery is a life-threatening situation for both mother and baby if untreated. Furthermore, there is the potential for damage to surrounding tissues of the mother such as muscle, bowel, or bladder during a difficult birth. No wonder the human birth process is described as 'labour'. In contrast, our primate cousins can give birth without any fuss whatsoever. The compromise then, is for the human baby to be born before it gets too big to negotiate the narrow birth canal, even if it is not yet fully developed. The solution is to

Pelvic outlet

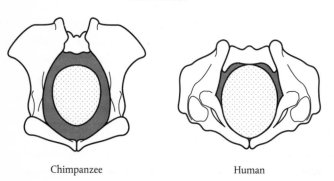

Chimpanzee   Human

FIGURE 6 Comparing the size of the pelvic outlet in relation to the fetal head in a female chimpanzee and a woman.

postpone much of the fetal growth to the postpartum period. This is why the human offspring is so altricial. It is estimated that in relation to its state of development, the human baby should ideally stay inside the womb for 21 months rather than nine before being born.

Fortunately Nature has endowed the mother with good postnatal care, a behaviour which is especially well developed in humans. As we shall see in subsequent chapters, genomic imprinting and placental hormones will ensure that the mother will engage in this important postpartum activity. Scientists now think that the incremental increase in human intelligence and brain size may have reached a plateau, and could even go into reverse if evolutionary pressures favour such a decline. If this were to happen, then easing the trauma of the human birth process would certainly be a major driving force.

Why have humans chosen to walk on two legs if it causes the birth process to be so difficult? All our ape ancestors who descended from trees continue to walk on four legs or at least are 'knuckle walkers'. Many biologists argue that bipedalism is the human characteristic most difficult to explain in evolutionary terms. The ability to walk upright has often been touted as one of the pivotal innovations that has made humans the dominant species on the planet. There are many theories, all of which are plausible but difficult to prove since whatever happened, happened a long time ago. Freeing the hands for complex tasks is probably the favourite. Then, standing tall is thought to have given humans an advantage in scanning a wider horizon in search of prey or to escape from predators. An erect posture will also reduce the surface area of the body that is exposed to the sun and hence increase range and stamina when hunting in the open. It has also been proposed that it is a form of sexual display. Stand up and dazzle the opposite sex with a full frontal seems to be the evolutionary message. But the cost is high.

The human birth process has become significantly more complicated compared to other mammals. To sacrifice reproductive efficiency for bipedalism does not seem a fair exchange in terms of survival of the species. Bipedalism might have evolved before humans started to develop a larger brain. When humans began to walk upright about 6 million years ago, brain size was much smaller, about a quarter of what it is today. There was no need to worry about birth difficulties then. But when brain size increased, it was too late to change back. Humans were already committed to walking on two legs and locked into the possession of a narrow pelvis.

Difficulty in childbirth is not a modern phenomenon. A 2,000 year old female skeleton was recently uncovered in South Africa with signs of disruption of the pubic symphysis. This is the joint where the right and left hip bones come together in front of the body. Her pelvis was unusually narrow so the injury could have been caused by birth trauma. Looking out for similar injuries in older specimens of pelvic bones might tell us when humans first experienced childbirth difficulty.

Because the human haemochorial placenta is so intrusive, the pregnant mother needs to have some degree of control. She solves this problem by forming a specialized lining in her uterus called 'decidua' (*deciduus* = falling off, shed).[28] In humans, the decidua is formed during the last part of the menstrual cycle even before pregnancy occurs. If there is no pregnancy, the decidua is shed through menstruation, but if the woman becomes pregnant the decidua is maintained. So decidua is formed not as a result of pregnancy but in anticipation of pregnancy. In most other species such as mice, decidua develops only after pregnancy has occurred and is triggered by the presence of the implanting embryo. We do not know why humans have switched from embryonic to maternal control of the process of decidualization but its consequence is clear. Women have to menstruate every month in order to rid the uterus of decidua and replace it all over again with the non-pregnant uterine lining; rather wasteful in material and energy.[29]

Only those species with haemochorial placentas form decidua while those with the epitheliochorial variety do not, implying that it is only needed when the placenta is invasive. We can interpret this in one of two ways. The first is to regard decidua as being protective to the mother by providing a restraining influence on the placenta so that it does not invade too deeply. The second is the opposing view that the decidua is a layer of 'fertile soil' which actually invites the placenta to invade. Either of these interpretations could explain why the invasive haemochorial placenta is associated with the formation of decidua. They are not mutually exclusive. Placental biologists have not quite made up their minds whether the point of contact between the fetal placenta and the maternal uterus is an area of conflict or cooperation. There is evidence to support both these views.

The restraining influence of decidua is illustrated very well in animal experiments where the placenta is transferred elsewhere in the body where there is no decidua. In these areas outside the uterus, placental invasion is totally out of control. In human pregnancy, when the placenta overlies areas deficient in decidua, such as a previous Caesarean section scar in the uterus or in ectopic tubal pregnancy, there is also excessive invasion. Normally the decidua is the site of placental separation from the uterine wall after the birth of the baby. Lack of decidua produces a condition called 'placenta accreta' where the placenta is so firmly stuck to the uterus that it cannot be removed at birth.[30] This is a serious pregnancy complication and is one of the major causes of postpartum haemorrhage, requiring removal of the uterus (hysterectomy) to stop the bleeding. Placenta accreta was first reported in the 1930s. There were no descriptions of this disease before that time. It is unlikely that such a serious condition remained undetected, suggesting placenta accreta is a disease that has newly arrived on the obstetric scene. Its incidence has increased steadily over the past 50 years because of the growing trend in births by Caesarean section leading to scarring of the uterine wall and deficient decidua formation. Middle-class women on the Indian subcontinent are reported to be opting for Caesarean sections so that their child can be born exactly on an auspicious day. There is also the growth of the 'too posh to push' culture in some societies that is increasing the popularity of Caesarean section. Prior Caesarean section is now recognized as the most important risk factor for placenta accreta, putting it into the category of an 'iatrogenic' disease, defined as a condition that is the unfortunate side effect of a clinical procedure. Should this surgical intervention in obstetric practice be discouraged?[31]

New guidelines from the National Institute for Health and Clinical Excellence (NICE) in the United Kingdom recently declared that all women can elect to have Caesarean sections on the NHS even if there is no medical need. This immediately provoked a heated debate between the pro-lobby and anti-lobby. The former claims this will empower women to choose the way they give birth while the latter thinks it a terrible idea to offer surgery to people who are not ill. I am particularly amused by the disdainful description in a daily paper of the anti-lobbyists as a bunch of 'placenta-munching' individuals, obviously referring to the fact that this habit is not for the modern, sophisticated woman. I would have thought it should be the other way round; that it is the more enlightened woman, freed from the constraints of conventional taboo and well-informed about the nutritious value of the placenta, who are most likely to 'munch' on the placenta. Still, all this offers a new take on attitudes towards placentophagy (see Chapter 9).

Even in normal pregnancy when the placenta has implanted over a layer of decidua, separation of the placenta at birth is invariably accompanied by serious blood loss. This contrasts with animal species with epitheliochorial placentation where there is hardly any blood spilt at parturition, such as in calving, foaling, or lambing because their superficially attached placentas peel off readily. This easy unwrapping is not available to the deeply invasive haemochorial human placenta.

As we shall see in subsequent chapters, there are many components in decidua such as NK cells, cytokines, and adhesion molecules which can facilitate or hinder the movement of placental cells. Decidua, therefore, can exert a positive as well as a negative control on how deeply the placenta is permitted to implant. In this way, the mother has ultimate control over how the placenta should behave. The line of demarcation between placenta and uterine wall is where two worlds meet. It is always the same in every pregnancy. The placenta penetrates as far as the inner third of the myometrium (muscle layer of the uterus) and no further. The decidua has been likened to a 'Procrustean bed' forcing conformity on the placenta. In Greek mythology, Procrustes (whose name means he who stretches) was a host who adjusted his guests to their beds. If they were longer than the bed, he cut off their limbs to make them fit; if too short, he stretched them to fill the space.

There is now an additional function proposed for the decidua. This is that it acts as a sensor to monitor the quality of embryos that are about to implant and will allow only the healthy ones to do so. This selection is necessary because human embryos are frequently defective. Females of most animal species exhibit a period known as 'estrus' (on heat) which is accompanied by clear external signs, such as dramatic changes in the colour of the female's face and other parts of her body, that she is now receptive to a male. This period coincides with ovulation so that when mating takes place a fresh egg is guaranteed to be available for fertilization. Humans do not have estrus and external signs of ovulation, if any, (such as a small rise in body temperature) are very indistinct. Even the woman herself may be unaware that ovulation has taken place. The result is that mating is generally not restricted to the time of ovulation. Eggs are frequently fertilized when they are well past their sell-by date. These eggs are particularly prone to genetic abnormalities and need to be screened out by the decidua. This could be why, in humans, the decidua is formed during every cycle even if no pregnancy occurs, whereas in most other species decidualization only begins when an implanting embryo is present.

Many embryos are lost very early in human pregnancy, often even before the mother is aware she has conceived. This is good economic strategy. To lose a defective embryo early is less costly to the mother in terms of time and energy than to carry it for longer and then see it miscarry later. For decidua to test the viability of the embryo first before allowing it to progress further makes good sense. 'If it were done when 'tis done then, 'twere well it were done quickly, (Macbeth).' This does sum up what decidua tries to do.

3

The Principal Players

One summer's evening, five biology Fellows of King's College, Cambridge, happened to be dining together at High Table. During the course of dinner, we were surprised to learn that, although we came from diverse disciplines, what we had in common through all our research was a little known cell from the placenta called 'trophoblast'. This could not be just coincidence. Rather, it points to the remarkable versatility of trophoblast and the pivotal role it plays in biology. Two significant developments spun off from this chance dinner encounter. First, we succeeded in persuading the Novartis Foundation to host a meeting for us to discuss trophoblast. Under the banner of the Novartis Foundation Symposium, this giant pharmaceutical company (formerly Ciba before it merged with Sandoz) funds and organizes regular conferences in their London headquarters on subjects which are timely and of broad scientific interest. They assemble a group of like-minded people together to debate a particular topic in depth. An invaluable part of these meetings is the subsequent publication of the entire proceedings, including the details of the discussions after each session so that anyone who is not present but reading the volume afterwards has the feeling of actually having been there. Our own meeting spawned a volume entitled *The Biology and Pathology of Trophoblast* published by Cambridge University Press.[32]

The second, and perhaps more important, development arising from our dinner was the decision by Cambridge University to establish the Centre for Trophoblast Research (CTR) on the back of a generous endowment by an alumnus with a deep interest in placental biology.[33] This is the only centre in the world which is entirely devoted to the study of trophoblast. The aim of the CTR is to promote research on the placenta. It has a range of activities: it provides travel funds for young investigators to attend and present their work at international conferences; the CTR itself regularly organizes its own meetings; it supports studentships as well as more senior Fellowships; and most importantly it fosters interdisciplinary research across departmental boundaries to study the placenta. Cambridge, like most ancient universities, is organized into multiple discrete departments where both teaching and research tend to take place. The CTR tries to encourage these disparate groups to come together. For the first time, there is someone from the department of engineering looking at the haemodynamics of blood flow within the human placenta. Engineering science is not a discipline one traditionally associates with reproductive biology.

I describe the above two events in order to show that the trophoblast cell is considered to be sufficiently important to deserve serious high-level scientific support. And yet, the name would probably be unfamiliar to most members of the general public. So, what is trophoblast and what are its functions? This and subsequent chapters will explain. It is the cell that gives the placenta its unique characteristics. Another reason why I have recounted this story is to counter the view (held by some) that college High Table conversation is usually nothing more than gossip and tittle-tattle. Something useful does emerge from time to time, as seen on this occasion. Four of the original dining companions are still around and continue regularly to discuss trophoblast. Sadly, one is missing. Anne McLaren, our embryologist, was killed in a tragic motor car accident.

Until about the middle of the 19th century there was no serious study of how the human embryo is formed. At that time, religious belief in divine creation made any such study unnecessary. Indeed, it was frowned upon even to question how life began. Since then, increasing scientific curiosity and the availability of appropriate experimental tools have changed the picture. In particular, the invention of the microscope together with the introduction by Edwin Klebs in 1869 of the cutting of very thin tissue sections for histological examination led to rapid advances in embryology. Much of the initial literature on this subject was published in the German language during this era because that country was at the forefront of embryological research.

The first step towards formation of the placenta is taken soon after the egg is fertilized by a sperm. This starts a series of cell divisions within the fertilized egg, first into two cells, then four, then eight, 16, and so on until a solid mass of cells is formed (see Figure 7). This is the defining moment when life begins. The first cell division or cleavage starts very early, as the egg is passing down the fallopian tube between 24–30 hours after fertilization. Around the 16-cell stage, two discrete populations of cells are already distinguishable: a central core known as the 'inner cell mass' (ICM) because of its location, surrounded by a peripheral layer of cells on the outside termed the 'trophectoderm'. The ICM is destined to become the embryo proper while the trophectoderm gives rise to trophoblast cells that will eventually form the most important component of the future placenta. Remember the name trophoblast, for these cells are the main characters in the placental story.

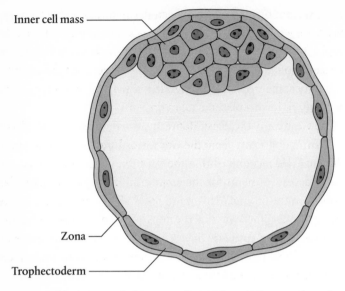

FIGURE 7 A blastocyst with the inner cell mass that will become the embryo and the trophectoderm that will give rise to trophoblast cells of the placenta.

This early separation of the trophectoderm from the ICM to produce a distinct trophoblast lineage is unique to mammals. Some of the genes that determine the initial split between the ICM and trophectoderm have now been identified.[34] Two genes known as *Nanog* and *Oct 4* are switched on very early in the ICM while another gene called *Cdx2* is the first to be activated in trophectoderm. These genes point out the developmental direction the initial primitive collection of cells should follow: either to become the ICM or the trophectoderm, that is, to be baby or placenta. In experiments with mice in which the *Cdx2* gene is artificially disrupted, no trophectoderm is formed, showing that the function of this gene is to drive the primitive collection of cells towards the formation of trophectoderm. In contrast, suppression of *Oct 4* leads to cells of the ICM differentiating into trophectoderm. The function of *Oct 4*, then, is to prevent the ICM from turning into trophectoderm during normal development. So, while *Cdx2* actively encourages the early cells to become trophectoderm, *Oct 4* discourages cells of the ICM from following the same route. It seems that the default mode is for the primitive cells to become trophectoderm unless something stops them from so doing. To form a placenta takes precedence.

The German pathologist Hubrecht in 1889 was the first to coin the term 'trophoblast' from the Greek word '*trophos*', meaning nourishment.[35] He recognized that the function of these placental cells was to access nutrients for the embryo from the uterus, like 'plant roots from the ground'. This was a novel idea at that time. The prevailing thinking then was that the fetus gained most of its sustenance from swallowing amniotic fluid (the 'waters') which surrounded it. The nutritional role of trophoblast is now well established, but this is but one of several functions. There are many more, as we shall see in subsequent chapters, so Hubrecht was only partly right.

By the fourth day after fertilization, the dividing egg has reached the uterine cavity. It now consists of approximately 55 cells, of which about five make up the ICM and 50 the trophectoderm, an astonishing difference in cell numbers. This unequal division clearly emphasizes where the priority lies. Almost all the energy is devoted to making cells for the placenta rather than for the embryo. In animals that lay eggs, nourishment of the embryo is available in the form of yolk. The mammalian egg contains very little yolk so an alternative means to feed the rapidly growing embryo must be found quickly. This is a job for the trophoblast cells and is why a large proportion of the cells formed after fertilization is immediately allocated for this purpose. This early separation of trophectoderm from ICM also means that the embryo and the placenta can now pursue their own programmes of development independent of each other, which is why trophoblast cells have different characteristics from embryonic cells in spite of coming from the same fertilized egg and sharing the same genes. Even their lifespans are different. While the placenta survives for only about 270 days, the embryo continues its existence for three score and ten years and beyond. The placenta has a life of its own. If the embryo were to die in the womb the placenta can live on but not vice versa.

The elegant experiments of Janet Rossant in the early 1980s demonstrated convincingly the crucial importance of trophectoderm in early pregnancy. Using two strains of mice, *Mus caroli* and *Mus musculus*, she transferred fertilized eggs of *Mus caroli* into *Mus musculus* females. There were no live births, which was not surprising since they were different strains. However, if the fertilized eggs were constructed so that the *Mus caroli* ICM was covered by *Mus musculus* trophectoderm and then transferred to *Mus musculus* recipients, successful pregnancies were achieved. What these experiments show is that the trophectoderm, and not the ICM, determines whether a fertilized egg is accepted by the mother. Many human embryos are lost early in pregnancy. Might this be caused by an incompatibility between trophectoderm and maternal uterus, resulting in failure to implant? We will look into this in Chapter 6. The trophectoderm already determines the outcome of the pregnancy even before the placenta itself is fully formed.

The above experiments lead us into the longstanding question of why there are different species. 'In the whole of biology there can be no concept that is at once so fascinating, so simple in outline and yet so baffling in detail, so much discussed and yet so little understood as the origin of species' remarked Roger Short in *Reproduction in Mammals*. If closely related species were able to cross-fertilize freely, they would gradually merge into one and distinct species would no longer exist. But inbreeding between different species is uncommon under natural conditions. What keeps disparate species from breeding successfully with each other? There are innumerable obstacles strewn throughout all stages of reproduction. Starting with mating opportunities or the lack of them, geographical isolation is one factor that immediately springs to mind. If you don't meet, you don't mate. Among species living in the same area, mating preferences can keep them from interbreeding. Some may breed in spring while others prefer winter, so the timing will never coincide. The absence of mutual sexual attraction can also put a brake on any further progress to the courtship ritual. A peacock's flamboyant tail feathers might be appreciated by a peahen but could be regarded as an embarrassing vulgar display by the female of another species.

In the event that all the above obstacles are successfully negotiated and mating is accomplished, there is still the next set of impediments that can block embryonic development. As we have just described, a layer of incompatible trophectoderm is likely to be a serious hindrance to hybridization. At a later stage, after the placenta has formed, there might also be a hormonal obstacle. For example, sheep and goats do not interbreed. This is because the sheep placenta produces all the hormones necessary to maintain pregnancy whereas the goat placenta does not (see Chapter 7). Whose placenta would be used in the hybrid? There is a definite misalliance here. Nature has its own way to sort the sheep from the goats. Another crucial hurdle to hybridization is the extensive variability in types of placentas found in mammals. The combined immunological, endocrinological, and structural characteristics of the placenta present a formidable reproductive barrier to interspecies breeding. This should come as no surprise since the placenta holds the key to successful pregnancy even within the same species.

Some cross-breeds do succeed in producing viable offspring but they have bleak futures. The mule, a cross between a female horse and a male donkey, is probably the best known example. The mule has been bred as a working animal since ancient times. Homer (800 BC) described it in the *Iliad*. But the mule suffers from one major defect: it is sterile because it has inherited an odd number of chromosomes from its parents. While the horse has 64 chromosomes and the donkey 62, the mule is in between with 63. No progeny can result from this odd number. The mule has come to an evolutionary dead-end because it can no longer reproduce its own kind. Tomorrow's mules will always be dependent on repeated crossing of today's horses and donkeys.

The mule is born from a donkey father and a horse mother. When mule breeders reversed the hybridization and mated a male horse with a female donkey, they found to their amazement that the resultant progeny was no longer a mule but what we now call a hinny. While the mule is larger and stronger and looks like an oversized donkey, the hinny is smaller and more horse-like. Even their temperaments are different. The mule is stubborn and belligerent while the hinny is more passive and docile. Why are mules and hinnies so different? After all, they have the same combination of parents and have inherited the same mixture of genes. This had been a puzzle ever since mule-breeding began thousands of years ago. It is only recently that an answer has appeared. This is related to a newly recognized form of inheritance called 'genomic imprinting'. Genes inherited from the father and those from the mother behave differently, a phenomenon geneticists describe as the 'parent-of-origin effect'. This could explain why mules and hinnies differ depending on the lineages of their parents. Who is the father and who is the mother is important. Other kinds of hybrids also follow the same rules. For example, a cross between a male zebra and a female horse results in a 'zorse' while the reverse mating of a male horse and a female zebra produces a 'horbra'. A zorse and a horbra are different like a mule and hinny. Similarly, a male tiger and a female lion gives rise to a 'tigon' while a male lion mated with a female tiger will produce a 'liger'. Again, a tigon and a liger have different characteristics.

Biologists are now beginning to appreciate that genomic imprinting and the parent-of-origin mode of inheritance is not just a convenient explanation for hybrid differences but is crucial to the whole of mammalian development, especially for the placenta. Its influence is so critical that we shall devote the whole of Chapter 4 to the subject.

The early trophectoderm is already sufficiently specialized to transport fluid and important chemicals into the egg to surround the ICM, creating a fitting microenvironment for the growing embryo. This fluid-filled structure is now known as a 'blastocyst'. Compared to the trophectoderm, cells of the ICM remain relatively primitive and are still not yet committed to form any particular type of tissue, nor have any specialized function. In other words, they remain pluripotent. Because of their pluripotency, cells from early embryos are the best source for harvesting stem cells but, for this, embryos have to be sacrificed. Herein lies the ethical dilemma.

Within the next 24–36 hours in the uterine cavity the number of trophoblast cells in the egg has almost doubled from around 50 to 100 cells. The rapid growth of trophoblast at this stage is truly impressive. And yet attempts to propagate these cells for use in the laboratory have been largely unsuccessful.[36] A variety of different growth recipes and techniques have been tried but to no avail. Trophoblast cells have stubbornly refused to cooperate. I suspect we are too late to catch those trophoblast cells which are still capable of dividing. The youngest placental specimens available from elective terminations that can be used for research are 5–6 weeks gestation age. By this time the proliferative phase for trophoblast may well have passed. They have stopped dividing and moved on to become invasive, which is their next function. Trophoblast mobility and proliferation are mutually exclusive. It may be necessary to backtrack to an earlier stage of development and harvest trophoblast cells from the blastocyst, but this would mean facing innumerable practical and ethical problems.

This is very frustrating for investigators wanting to use trophoblast cells in their research. At present a fresh batch of trophoblast cells have to be harvested from a newly collected placenta every time an experiment is planned. The procedure has to be started immediately after the placenta is collected, otherwise the trophoblast cells will die before they can be put into culture. The isolation process itself takes several hours and the cells collected have to be nursed in an incubator for another 2–3 days before use. Frequently a placenta becomes available at odd hours. What a luxury it would be if a supply of trophoblast cells could be constantly available in the laboratory ready for use. No wonder, then, that a great deal of research effort is nowadays directed at attempts to create a trophoblast cell line that can be propagated continuously in the laboratory. The Holy Grail is to identify trophoblast stem cells from which these trophoblast cell lines can be initiated. Mouse trophoblast stem cells have now been isolated[37] but, so far, the human equivalents have remained elusive.[38] Not only will a trophoblast cell line be more convenient, it will make results from experiments more valid. If all trophoblast cells used in every experiment originated from the same source, it would be much easier to extrapolate observations across multiple experiments. Unfortunately, at the time of writing, a genuine normal human trophoblast cell line is not yet available in spite of claims to the contrary. A persistent difficulty is how to recognize a trophoblast cell *in vitro* because cells cultured in the laboratory all look very much alike. What parameters does a cell have to satisfy before it can be defined as trophoblast? This issue is a major source of conflict. Some argue that trophoblast cells *in vitro* may lose the characteristics they possess *in vivo* leading to a situation in which some cell cultures may indeed contain trophoblast but there is no way to prove that they do: a Catch-22 situation.

There is an alternative. Three trophoblast cell lines are presently available—named Bewo, Jeg, and Jar—but they have all originated from a cancerous growth of the placenta known as a choriocarcinoma. They are not from a normal placenta so this is an important caveat. These cells are stable and possess characteristics of normal trophoblast, including the ability to secrete placental hormones, although they do have an abnormal number of chromosomes (86 instead of the 46 of a normal human cell). Roy Hertz, working at the United States National Cancer Institute, started it all in 1959 when he transplanted pieces of choriocarcinoma from a patient into the cheek pouches of Syrian hamsters where they remained viable for many years.[39] The hamster cheek pouch is described as an 'immuno-logically privileged' site because it allows foreign tissues to grow inside it without triggering rejection. Other similar sites are the cornea, the anterior chamber of the eye, and the brain. The cheek pouch has no lymphatic drainage, which is the conduit an organ uses for transporting foreign molecules that have gained entry to the body to alert the host's immune system. Without this signal, the hamster is blissfully unaware of the presence of the cancer tissue inside its pouch, allowing it to grow unimpeded. At one time even the uterus was regarded as a privileged site to explain why the fetus is not rejected by the mother, but this idea is now no longer tenable (see Chapter 6). Using the pouch tissue, in 1968 Roland Pattillo and George Gey succeeded in establishing the cells in culture. This was the first ever continuous human, hormone-secreting trophoblast cell line to be created in the laboratory.[40] These cells are still growing today and are used by researchers all over the world. They have attained a state of immortality. Although we have derived a great deal of information from the use of these cells, there is a lingering suspicion that they may not be truly representative of how normal trophoblast behaves.

Choriocarcinoma is a very unusual tumour with features not found in other kinds of cancer, so both normal trophoblast and its abnormal counterpart are equally fascinating.[41] Choriocarcinoma has an unusual, gross appearance and its histological structure is bizarre. At one time, its very origin was hotly debated. What kind of cancer is it? What tissue did it arise from? It was Felix Marchand, a pathologist from Marburg, Germany, who in 1894 initially established the trophoblast origin of choriocarcinoma. Although March-and's view was readily accepted in continental Europe and the United States, there was a series of heated and acrimonious debates in the Obstetrical Society of London whose members obstinately refused to believe him. The controversy was finally settled when, in 1929, Bernhard Zondek demonstrated the secretion of the placental hormone HCG (see Chapter 7) by choriocarcinoma.

My first encounter with choriocarcinoma was in Malaysia in the early 1960s. As a junior research assistant newly qualified from the United Kingdom, I was given the task of compiling the cancer register at the Institute for Medical Research in Kuala Lumpur. Many cancers have distinctive patterns of geographical distribution, being more common in some countries while rare in others. Choriocarcinoma is one of these, with a high incidence in Southeast Asian countries, so it featured prominently in the Malaysian cancer database. Why this cancer should occur more often in that part of the world is still unclear. In Chapter 4, we will look at another tumour of the placenta called hydatidiform mole, which has arisen as a result of disruption of imprinting. This tumour is also common in Asia. While choriocarcinoma can originate from a normal placenta, it occurs 1,000 times more frequently following a hydatidiform mole. The two are linked in some way. After spending some time delving into the mysteries of choriocarcinoma, I came to realize there were many unanswered questions relating to the normal placenta. Thus began my life-long research interest in reproductive biology. My career, therefore, followed an unusual path—backwards as it were—from studying the abnormal to the normal. As I shall now describe, the normal placenta and choriocarcinoma are but two ends of a spectrum. It doesn't matter at which end one starts.

Medical students during their pathology course are indoctrinated with the three cardinal signs by which a malignant or cancerous tumour can be distinguished from a benign one. These signs are cellular pleiomorphism, invasiveness, and metastases. With most cancers these are useful diagnostic features, but not for choriocarcinoma. Take cellular pleiomorphism. This means cells of irregular shapes and sizes. The presence of these cells suggests that they are derived from a cancerous growth and, in general, the greater the degree of cellular pleiomorphism the more aggressive the cancer is likely to be. This is the basis of the cervical smear screening test for cancer of the cervix, sometimes also referred to as the PAP smear because it was originally devised by Georgios Papanicolaou in the 1940s. Scrapings are taken from the surface of the cervix and examined for the presence of abnormal looking cells. The criterion of cellular pleiomorphism falls down when it comes to choriocarcinoma. At the normal implantation site where the placenta contacts the uterus, trophoblast cells of all shapes and sizes are to be found, ranging from unicellular to multicellular ones. Furthermore, trophoblast cells are intimately mixed with uterine cells with their own characteristics, so that when examined under a microscope the whole area has often been described as resembling a battlefield. A similar picture is seen in choriocarcinoma. The histological distinction between normal and abnormal is, therefore, blurred. Normal pregnancy has occasionally been misdiagnosed as choriocarcinoma and cases of choriocarcinoma have been missed as normal pregnancy, even by experts. The pathologist Wallace Park had a pithy comment: 'If the patient died she had choriocarcinoma. If she survived she did not'!

Invasiveness, the second cardinal sign of malignancy, is also inadequate to distinguish normal from abnormal trophoblast. Cells and tissues of bodily organs are normally placed in well-defined compartments so that one does not intrude into the other. The architecture remains constant. Take the cervix again as an example, its surface epithelial lining does not normally encroach into the underlying tissue. If it does, then this is evidence that malignant change has occurred. Following the detection of abnormal cells in a cervical smear the next stage in the screening process is to take a small piece of tissue from the cervix (biopsy) and examine it histologically for evidence of invasiveness. These are the steps taken to diagnose cervical cancer. In the case of the placenta, invasion by trophoblast cells through the uterine epithelium into the underlying tissue, even destroying the blood vessels, is a normal event. This is what happens during normal implantation (see Chapter 5). Without implantation the embryo cannot develop any further. Invasiveness then is not diagnostic of choriocarcinoma. Even normal trophoblast is highly invasive.

We now come to metastasis, the supreme hallmark of a malignant tumour. This is when cells break off from a primary growth and are disseminated to other organs throughout the body. This is how cancer kills the patient and is why doctors try to detect the presence of a tumour early while it is still localized before it has spread widely. Unfortunately, in many cancers, particularly those of internal organs, metastatic deposits provide the first signs that anything is wrong. By then it is usually too late. Cancer of the ovary is a prime example. It is usually silent and symptomless until massive spread has occurred. Surprisingly, we find normal trophoblast exhibiting something very similar to cancer metastasis. In every normal pregnancy, trophoblast cells constantly break off from the placenta, enter the mother's bloodstream and are carried off to lodge in her lungs in the same way as choriocarcinoma would do.[42] The lungs of most pregnant women (perhaps all) contain fragments of trophoblast, known as syncytial sprouts, which have broken off from the surface of the placenta and become lodged in the lungs. They remain there for a while and then disappear, presumably destroyed by the mother. It is estimated that as many as 100,000 sprouts enter the maternal circulation daily during pregnancy. This trophoblast 'deportation' as it is called, occurs only in humans and is not seen even in higher primates. There is much speculation about the physiological role of this phenomenon. Why does it occur? Is there a purpose to it? What do the sprouts do? Some people believe that this 'deportation' of trophoblast sprouts alerts the mother to the presence of her placenta, but why she needs this warning or how she is supposed to respond to this cryptic message is not clear. In the late 19th century, the German scientist Georg Schmorl (1896) noticed there were many more sprouts deposited in the lungs of women who had died of pre-eclampsia than in the lungs of normal pregnant women.[43] From this observation, he speculated that these clumps of trophoblast might release some substances which were 'toxic' to the mother. Thus, Schmorl was the first person to direct attention towards the placenta as the ultimate cause of pre-eclampsia. This is a major disease of pregnancy. We shall look at it more closely in Chapters 5 and 6.

In contrast to the situation in normal pregnancy, deposits of choriocarcinoma, after lodging in the lungs, do not disappear. They continue to grow and eventually spread further afield. Without treatment, choriocarcinoma will inevitably kill the patient. In the lungs, choriocarcinoma erode the blood vessels, creating massive areas of haemorrhage which is characteristic of this cancer. As we shall see in Chapter 5, the ability to erode blood vessels is also a property normal trophoblast uses during the initial stages of implantation. Choriocarcinoma relies entirely on invasion into maternal blood vessels for its nutrition, again mimicking the activity of normal trophoblast in early implantation. This is a fundamental difference between choriocarcinoma and other solid cancers. There are no blood vessels supplying choriocarcinoma. In other types of cancer, a network of new blood vessels generated by the cancer itself is a pre-requisite for its survival. Otherwise, the cancer will wither away from lack of food. Choriocarcinoma does not need new blood vessels. It is already bathed in a pool of maternal blood.

The most intriguing aspect of choriocarcinoma is that it has arisen from trophoblast and is, therefore, fetal in origin. It is the only example of a human cancer that has come from another individual—the baby. All other cancers start off from the patient's own tissues. We cannot 'catch' cancer from other people. We must not confuse this with the transfer of a cancer-causing agent, such as a virus. The Human Papilloma Virus (HPV) which causes cervical cancer *can* be passed on from one individual to another by sexual intercourse, but this is transmission of the virus not the cancer itself. Cancer has often been referred to as 'the enemy within'. But choriocarcinoma is the exception. It is not from within. It is derived from the fetal placenta. And yet, it manages to establish itself in a pregnant mother in spite of being from a 'foreign' tissue. This makes choriocarcinoma unique among human cancers and could explain its many unusual features.

A case has been reported in which a transplanted kidney contained deposits of choriocarcinoma which were undetected. Like all transplant recipients, the patient was given immunosuppressive drugs to prevent rejection of the donated kidney. Because the patient's immune response was damped down, the deposits of choriocarcinoma started to grow and spread. When the doctors realized what had happened, the immunosuppression was stopped. With the return of immunocompetence, the cancer regressed and was eliminated. This accidental human experiment demonstrates clearly that choriocarcinoma cannot grow in a totally unrelated host whose immune system is intact. It therefore behaves like any other kind of cancer. And yet, choriocarcinoma survives in a pregnant woman in spite of having originated from fetal trophoblast. Is it because choriocarcinoma is half related to the mother since she has contributed, with the father, to producing the placenta from which the tumour is derived? Is a pregnant woman immunosuppressed? These are intriguing questions. We shall be looking more closely at the immunology of normal trophoblast in Chapter 6. Much of the information presented there will also be relevant to choriocarcinoma.

Transmissible cancers are also extremely rare in animals, with only two documented examples: one is the venereal cancer of dogs spread by sexual intercourse and the other is the very aggressive cancer in Tasmanian devils transferred by biting. The Tasmanian devil is a rare carnivorous marsupial whereas most other marsupials, such as the kangaroo, are herbivores. They were eradicated in the Australian mainland by new predators like the dingo, introduced by early settlers, but have found refuge in Tasmania. Now their survival is threatened and they are on the verge of extinction so scientists are making a concerted effort to try and understand why their tumour behaves in this unusual way, a conundrum shared by choriocarcinoma. A popular theory is that the devil's prolonged isolation in Tasmania has led to excessive inbreeding with the result that there is little genetic variability between individual animals. This allows the cancer to be freely disseminated within the population. Recent genomic sequencing of devils from different parts of the island confirms that their genes are indeed not too different from each other. In spite of this high degree of relatedness, the animals are still able to reject skin grafts exchanged between them, indicating that their immune system is intact. And yet, both choriocarcinoma and the Tasmanian devil's cancer have managed to evade this immune surveillance of their respective hosts.

The best news about choriocarcinoma is that it is the only solid cancer that is curable. Once, choriocarcinoma was one of the most aggressive cancers. Almost all patients died within a year of diagnosis. In the early 1940s, Roy Hertz noticed that embryonic tissues needed folic acid to survive and that folic acid antagonists can cause miscarriage in women. This observation led to the first treatment of choriocarcinoma patients in 1956 with the anti-folic acid drug methotrexate. The result was dramatic. The introduction of methotrexate had completely transformed the outcome. Almost 100 per cent of patients, even those with widespread disease, can now be cured. This is a real success story in cancer drug therapy. In no other solid cancer has such a reduction in mortality been achieved. The outcome was so unexpected that clinicians did not dare to use the term 'cure' for some time. Instead, it was referred to merely as 'sustained remission' leaving the underlying suspicion that one day the cancer would return. But cancer did not reappear in any of the treated cases. It was only then that the chemotherapy was accepted as completely successful. It is still the general rule today that widely spread

cancer remains incurable, although life can be prolonged to a varying extent. A five-year remission is all that can be hoped for. Therefore, in its clinical behaviour, choriocarcinoma is quite unlike any
other human cancers. This may be because choriocarcinoma is of
'foreign' origin (from trophoblast) and, hence, potentially susceptible to immune rejection. The immune response on its own may not
be sufficiently robust to get rid of the cancer but, in conjunction
with chemotherapy, the two together will tilt the balance in favour
of the patient. Choriocarcinoma can sometimes arise in males from
the pluripotent germ cells of the testis. In this situation, the choriocarcinoma is, of course, from the patient's own cells, unlike choriocarcinoma from the placenta in a woman following pregnancy.
Testicular choriocarcinoma is relatively resistant to chemotherapy,
which seems to support the importance of the immune response in
eliminating placental choriocarcinoma.

Throughout this book the similarity between trophoblast and
cancer is constantly emphasized. Nowhere is this resemblance more
evident than that seen between normal trophoblast and its malignant counterpart, choriocarcinoma. Indeed, the normal behaviour
of one and the abnormal misbehaviour of the other merge to such
an extent that, as I mentioned earlier, they can be said to form a
continuum. Where is the line to be drawn? Take ectopic or tubal
pregnancy. Here we have a normal placenta implanted in the wrong
place. If allowed to develop, it will kill the mother as any malignant
growth would do, albeit by a different mechanism. Wallace Park
summed it up very well: 'Choriocarcinoma is not pregnancy un-
checked' nor 'successful pregnancy a failed choriocarcinoma'.

Enough about abnormal trophoblast. Let us now turn our attention back to normal trophoblast. At first the trophectoderm surrounding the blastocyst consists of a single layer of cells but soon
their repeated proliferation creates three different populations of trophoblast. These are **syncytiotrophoblast**, **endovascular trophoblast**,
and **interstitial trophoblast**.[44] They have diverse functions

and come into contact with different areas of the uterus. They all contribute in their own way to make placentation successful, as will become clear in Chapters 5 and 6.

Syncytiotrophoblast (see Figure 8) is extraordinary in that it is in the form of a syncytium, meaning a confluent sheet with multiple nuclei but no intercellular boundaries, like a long string of Cumberland sausage without any kinks in between. No other cell in the

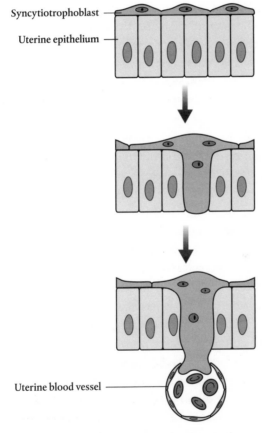

Syncytiotrophoblast

Uterine epithelium

Uterine blood vessel

FIGURE 8 Syncytiotrophoblast invading through the uterine epithelium to make contact with uterine blood.

body exists as a syncytium. Syncytiotrophoblast penetrates the uterine epithelium to start the process of implantation and will form the lining separating the placenta from the mother's blood. In 1882 the German anatomist, Theodor Langhans, was the first to recognize syncytiotrophoblast as part of the placenta. Previously, this layer was thought to be derived from the mother because there was a reluctance at that time to accept the revolutionary idea of a fetal tissue being bathed directly in maternal blood. It was easier to imagine the syncytiotrophoblast as the wall of some enormously dilated maternal blood vessel than part of the fetal placenta. The reason why blood in the circulation does not normally clot is because the inside of blood vessels is lined by a special layer, the endothelium, which has special non-clotting properties much like the Teflon coating of frying pans. The surface of syncytiotrophoblast is endowed with similar features. It also has a high electrical negative charge, which might explain its non-stick quality. When two surfaces of similar charge come into contact, they tend to repel each other. Red blood cells too have a high surface negative charge, which is why they don't clump together. All these features allow maternal blood to flow freely within the placenta.

Syncytiotrophoblast will eventually come to line the entire surface of the placenta, separating it from the mother's blood. It is responsible for accessing nutrients from the mother and is also the principal source of all the hormones synthesized by the placenta. From its position, syncytiotrophoblast has direct access to the maternal circulation. The job of the interstitial trophoblast subset is to alter the calibre of the uterine arteries, a process that is critical for creating an adequate blood supply for the developing embryo, while the endovascular trophoblast forms a plug over the mouths of the arteries to prevent maternal blood from coming in at too high a pressure. All this will become clear in Chapter 5 on implantation. The mature placenta, then, is made up of several trophoblast subpopulations with dissimilar features, distinct functions, and which contact either maternal blood or uterine tissues. The placental–maternal interface is not uniform. It is made up of disparate cell types on both sides.

The surface of syncytiotrophoblast has two further exceptional characteristics that fit well with its function.[45] Using electron-microscopy, the surface of syncytiotrophoblast is seen to be covered with large numbers of hair-like projections called 'microvilli', like a fine brush. This increases enormously the surface area available for absorption of nutrients. Similar 'brush borders' are found in intestinal and renal epithelium, areas which are also involved in absorptive functions. If maternal nutrients are restricted or inadequate there is an increased proliferation of these surface microvilli, which is a compensatory change to enable the syncytiotrophoblast to absorb more from the mother.

All cells in the body have surface molecules inherited from our parents that are unique to each individual. These are called HLA (see Chapter 6). Apart from identical twins, no two individuals will have the same combination of HLA. This is the reason why organ grafts such as kidney or heart are rejected because the recipient's immune system recognizes the donated organ as not being his or her own. Syncytiotrophoblast is alone in being the only cell in the body which is totally devoid of HLA molecules. In this way it has devised a clever camouflage to make itself invisible to the mother's immune system. This is an important mechanism for placental survival *in utero*.

The syncytiotrophoblast is a truly remarkable cell. Not only is it unique in being the only cell in the body which exists in the form of a syncytium, its formation is itself a story within a story. This brings us to the intriguing tale of HERV (Human Endogenous Retrovirus). Since the dawn of time, humans have regularly been infected by viruses just as we are today. Many cause serious diseases like flu, measles, chicken pox, polio, and AIDS. Viruses do not have the necessary machinery to replicate themselves. They can only do so by 'borrowing' the genetic mechanism of the host cell they have infected. For this reason, viruses have to insert their own genes into the genes of the host in order to survive. Viruses, therefore, cannot exist independently. Once new viruses are formed, they come out of the cell to infect other cells and, ultimately, other individuals. This passage from person to person is known as 'horizontal transmission' and is the usual way viruses spread. But sometimes the virus can be passed on to the next generation via infection of the germ cells (sperm or egg), described as 'vertical transmission'. Many of these viruses eventually become 'defective', that is, they can no longer replicate themselves and assemble new viral particles but their genes are not lost, remaining integrated among the genes of the host cell. These viral genes are described as having been 'endogenized' in the host cell. Viral genes are made up of either DNA or RNA. In mammalian molecular genetics, the central dogma is that everything begins from DNA, progressing to RNA, which eventually makes the proteins needed for the cell. DNA viruses simply insert their genes into the DNA of the host cell they have infected. But for RNA viruses to get the host cell machinery going, they first of all have to insert their RNA genes *backwards* into host cell DNA using a special enzyme called 'reverse transcriptase'. These viruses are, therefore, known as 'retroviruses' (Greek: *retro* = backwards). Most endogenous viruses are retroviruses, hence the term Endogenous Retrovirus (ERV) with the human equivalent labelled as HERV.

One of the first hints that HERVs might be present in the placenta was in the early 1970s when the virologist Sy Kalter reported that he had detected by electron-microscopy tiny objects in a baboon placenta that looked like retroviral particles.[46] This was really a case of serendipity because the baboons were part of a group infected with rubella, the German measles virus, to study the congenital effects of this virus. The retroviral particles were a chance observation. The experiments were not set up to search for them. After this, similar particles were seen in the rhesus monkey, and in human as well as in mouse placentas, so this characteristic is not restricted to primates.

There is other indirect evidence. For example, antibodies made in the laboratory to identify cells infected with a retrovirus also react with syncytiotrophoblast. Conversely, antibodies made towards syncytiotrophoblast will react with retroviral-infected cells.[47] There is no reactivity between syncytiotrophoblast and cells infected with other types of viruses. This is an important control often used in scientific observations. It confirms that the cross-reactivity specifically involves retroviruses and is not just a general phenomenon of all viral infections. These observations show that the cell surface of syncytiotrophoblast and retroviral-infected cells share a similar profile, which leads to the obvious question: are the molecules on the surface of syncytiotrophoblast produced by some unknown retrovirus hidden inside the cell? Even more intriguing is the finding that blood from normal pregnant women will react against cells infected with retrovirus, suggesting that pregnant women actually recognize the presence of these retroviruses.

When the human genome was first analysed, scientists were astonished to find that approximately 5 per cent of the human genome is made up of HERV genes. These genes can be traced back as far as the early days of primate evolution, over 60 million years ago. We are, therefore, part human, part virus! Where have these viral genes come from? They probably represent past infections by retroviruses which became endogenized in the human genome and were then transmitted through subsequent generations right up to the present time, with the largest number having accumulated about 30 million years ago after separation of the Old and New World monkeys.

What are the functions of these HERV genes, if any?[48] As already mentioned, most HERVs are defective and are unable to make complete new viruses. They are, therefore, no longer infective to other individuals. But some continue to produce viral proteins. One such protein of increasing interest to placental biologists is 'syncytin'.[49] As its name implies, this protein has the ability to fuse cells, transforming them into a syncytium. This can be demonstrated in the laboratory. In the human placenta two variants of syncytin (syncytin-1 and -2) are found in great abundance. Similar syncytin proteins are also found in mouse and sheep placentas. Indeed, the ERV responsible for the production of syncytin is present in all eutherian mammals. Mice made deficient in the production of syncytin in the laboratory have impaired placental transport and exhibit defects in the labyrinthine area of the placenta (equivalent to human syncytiotrophoblast). The evidence is compelling that syncytin protein produced by HERV is the primary agent that induces syncytium formation in humans and other mammals. Besides syncytin, other HERV proteins have known immunosuppressive properties which could be important in protecting the placenta from immunological attack by the mother (see Chapter 6). Surprisingly, both syncytin-1 and -2 have recently been found in the brain and also in some other fetal organs besides the placenta. What their functions are in these other organs is unknown. HERV proteins may have more far-reaching effects on human development than we realize.

Even those HERVs which do not make their own proteins could have a role to play in the development of the placenta. A typical retroviral structure is bounded by two regions at either end called Long Terminal Repeats (LTR). These are areas which normally regulate the activities of the HERV genes. If these LTRs happen to be inserted near some genes of the host cell, they can promote these host genes to become active, without the need for the HERV itself to be productive.[50] So the HERV can actually encourage the host cell to make proteins which otherwise would not be made. Such a scenario is thought to be involved in the production of a protein called 'pleiotropin' which is an important growth factor for the development of the placenta. This protein is also said to be responsible for the invasive properties of trophoblast.

The idea that the evolution of our placental form of reproduction originated from retroviral infections way back in our ancestral past might sound incredible, but placental biologists are giving it serious consideration. Nature is highly inventive and is full of surprises. In 1967, the cell biologist Lynn Margulis introduced the term *endosymbiosis*, which can be translated as 'living together inside a cell'. The theory proposes that, in the early history of life on Earth, the ancestors of present-day cells ingested small bacterial life forms which eventually became part of the cell itself. The green leaves of modern plants are made up of tiny granules inside their cells called 'chloroplasts'. These chloroplasts were originally not an integral part of plants but were, in fact, bacteria which had colonized primitive plants when they first appeared on Earth billions of years ago. These bacteria developed the ability for 'photosynthesis'. This is the process which converts carbon dioxide into oxygen in the presence of sunlight. Over the course of time, these bacteria, because they were so useful, were permanently incorporated into the plant cells and became chloroplasts. All plants now obtain energy by photosynthesis. The emergence of photosynthesizing green plants had a dramatic impact on the Earth's atmosphere. The resulting accumulation of oxygen in the air led to an explosion of new oxygen-breathing life forms, the so-called 'aerobic' organisms like ourselves gradually replacing the previous 'anaerobic' ones. Another benefit of oxygen is that it led to the formation of the ozone layer which has protected the Earth from fierce UV radiation. This allowed new species to emerge from the oceans to colonize land.

To make the transition from anaerobic to aerobic, a mechanism must first be put in place that utilizes oxygen as a form of respiration. This brings us to another example of endosymbiosis. In the cytoplasm of all animal cells today, there are small particles called 'mitochondria'. These are the generators that convert oxygen into energy for the cell. Again, like chloroplasts in plants, mitochondria were once oxygen-breathing bacteria that colonized primitive organisms billions of years ago in the early days of the formation of our planet. They were incorporated into animal cells and eventually became permanent fixtures. If we look closely at the mitochondria in our cells, they still retain many features of their ancestral bacterial origin. The ancient bacteria from which mitochondria are derived belonged to a family known as Rickettsiae. These bacteria are still with us today, causing diseases such as tick-borne typhus. The interesting thing about Rickettsiae is that they are, and presumably were, obligate intracellular organisms; they can only live inside cells. Their search for a suitable habitat might have been what started their colonization of our ancestral cells.

There are innumerable other examples of symbiotic existence that can be cited. The colour of corals inhabiting the sea bed is due to algae living inside them. Otherwise, corals are colourless. The ability of the diphtheria bacterium to produce its life-threatening toxin is dependent on a virus within its genome. At first, the algae and virus were merely guests in the coral and bacterium but, after millions of years of living together, they became permanent residents.

With all these examples, the idea that HERV could be responsible for the formation of syncytiotrophoblast is no longer so improbable. The book *Virolution* by Frank Ryan on how viruses influence evolution is particularly persuasive. Still, it takes a lot of getting used to because in most people's minds viruses are invariably harmful. For example, the virus HIV that causes AIDS is a retrovirus. At present, it is transmitted horizontally from person to person to cause infection. In the distant future could it become integrated into our genome and become another HERV? If so, would it live within us in peaceful symbiosis like other HERVs instead of causing disease? It may even develop a useful function and play a role in the future evolution of our species. In a pathogen–host relationship, we frequently find that the most successful are those where the infection is not lethal to the host. No advantage is gained by the pathogen in killing the host. It then has to seek out a new host. By virtue of their vertical transmission, HERVs have solved this problem by being passed on from generation to generation. At the same time the host benefits by evolving a viviparous mode of reproduction—a perfect symbiotic relationship.

Evolution usually progresses slowly, step by step, as genes within organisms change or mutate over time. The ability to acquire whole new sets of genes from another source and use them as partners in an endosymbiotic relationship is an effective way to speed up the evolutionary process. Without colonization by HERV, would mammals have developed a placenta by simple genetic mutation alone?

4

Parental Tug-o-War

Since we are the product of our father and mother, we possess two copies of every gene, one inherited from each parent. Usually, both copies are active and contribute equally. But there is a small group of genes that do not behave in this way. Within this group, certain genes are active only when they come from the father, while others are so only when they are from the mother. This unusual parent-of-origin form of gene expression is due to a phenomenon known as 'genomic imprinting' and the genes that behave in this manner are called 'imprinted genes'. In scientific terminology, the active gene is said to be 'expressed', so imprinted genes are either paternal or maternal expressed depending on which parent they are derived from. Genomic imprinting controls whether certain information should come from the father or from the mother. All that is needed is to shut down either one or the other. What is the reason for this parental segregation?[51]

At the time of writing, at least 55 genes which are subject to imprinting have been identified in humans and 100 in the mouse. The rate of discovering new imprinted genes seems to be slowing so we may be reaching a plateau and it is unlikely that many more will be found. The numbers might even fall as some genes presently designated as imprinted are subsequently discovered not to be.[52] The human genome contains about 30,000–40,000 genes, so only a small minority are imprinted. Of the known imprinted genes, over half are found in the placenta of both humans and mice, suggesting they have a critical role to play in making the placenta.

In 1984, three ground-breaking papers were published in scientific journals describing the artificial creation of fertilized mouse eggs containing genes which are either exclusively from the father (androgenomes) or from the mother (gynogenomes), unlike normal fertilized eggs which have one set of genes from each parent.[53] Conceptuses born to androgenomes had well-developed placentas, while those born to gynogenomes had poorly developed placentas. These experiments gave the first hint that parental genetic contributions are not equal during placental development. Genes from the father provide the necessary information for growth of the placenta, while genes from the mother have the opposite effect. The two sets of parental genes balance each other to produce a placenta of optimum size. Because genes from both parents are needed for proper placental development, and hence successful pregnancy, parthenogenesis or self-fertilization of an egg without sperm does not normally occur in mammals. The Aristotelian view of pregnancy in the 4th century BC already alluded to the different input of the male and female parents. He likened the male to being the sculptor while the female provided the raw material. Together, they created a new life but it was the male who shaped the figure. Some feminist critics were not well pleased, labelling Aristotle a misogynist for proposing such a politically incorrect idea.

The first demonstration of parthenogenesis was by Jacques Loeb in 1899, a German scientist working in the United States who succeeded in stimulating unfertilized sea-urchin eggs to divide simply by treating them to different mixtures of salt solution. No sperm was needed. Self-fertilization was possible, at least in sea-urchins. Loeb was widely hailed as the modern day Faust, someone who could conjure up a new being from chemical processes. 'The Immaculate Conception Explained!' was a typical news headline following Loeb's achievement. In 1910 a French scientist, Eugène Bataillon, took Loeb's experiments a stage further from an invertebrate to an amphibian. He caused frogs' eggs to develop simply by pricking them with a needle. Again, no sperm was needed so this was another example of parthenogenesis. To Bataillon, the implication was clear. In natural conception, sperm merely acted as a mechanical stimulus to tickle the egg to start division. Their genetic contribution was thought to be unimportant. Some women welcomed this news for it made the male redundant if alternative means could be used as a substitute for sperm. This would allow women to control their own reproductive destiny.

Since egg-laying reptiles do not have a placenta and are not subject to imprinting, they can also reproduce by parthenogenesis. The giant lizard, the Komodo dragon, is a well-documented example. Between 2005 and 2008, several zoos in Britain and the United States reported the hatching of live offspring from unfertilized eggs by female dragons who had had no contact with males. Subsequent genetic testing confirmed there was no male contribution. Interestingly, all the offspring were male. The sex chromosomes of the dragon are Z and W, with ZZ being male and ZW being female. This contrasts with humans where males are XY and females XX. Thus, the heterogametic sex, that is the sex with the odd pair of sex chromosomes, is the male in humans but is the female in the dragon. The female dragon gives one of each chromosome to her eggs, either a Z or a W. When those eggs self-fertilize by parthenogenesis, the egg containing a Z will duplicate into ZZ which is a male, while the one with a W becomes WW which is not viable, in the same way that YY is not viable in humans. This explains why all the zoo dragon parthenogenetic offspring are male. Chickens and other birds also have Z and W sex chromosomes with cockerels being ZZ and hens ZW.

How Komodo dragons found themselves on one small island in the Indonesian archipelago had long been a mystery. With the discovery that they can reproduce by parthenogenesis we may now have the answer. An ancestral female dragon is thought to have been washed ashore on Komodo Island many million years ago. This single female castaway set about reproducing by parthenogenesis and succeeded in producing ZZ male offspring. She then mated with her own male offspring but, this time, the descendants could be either ZZ males or ZW females because they were no longer the products of parthenogenesis. From then on, a sexually reproducing population was established on the island. Of course, a human female castaway will not be able to emulate such a feat. Even if she were able to self-fertilize (which she can't), all her offspring will be XX females because the human egg carries only an X sex chromosome. The Y sex chromosome is in the sperm. Without male offspring a sexual reproductive model can never be restored. She will be condemned to a perpetual cycle of parthenogenesis. Self-fertilization then is only beneficial as an interim measure. Prolonged usage would doom the species to extinction.

Even those species that can reproduce by parthenogenesis have to revert to occasional bouts of sex in order to introduce new genes into the population, otherwise the species will die out, overburdened by the accumulation of bad genes that are not replaced. This is where the male of the species comes to the rescue. His sperm distributes genes between females. Without this constant introduction of new genes, females would eventually degenerate into parthenogenetic clones. Sperms are particularly susceptible to genetic change, with new genes springing up all the time. This is because sperms divide many times before becoming mature. With each division there is a chance that error is introduced when the genes are put back together again and some are altered from what they were before. In this way, new genes are formed. These new genes can be 'good' or 'bad'. Nature will decide. The good ones that are beneficial for survival will be conserved to spread through subsequent generations while the bad ones will be weeded out. This is how sexual reproduction speeds up evolution. It produces new genes for natural selection to act on.

The day when a male is no longer needed in reproduction is still some way off! This news should come as a welcome relief to all males of the species, especially as the media is full of sensational stories about the 'end of men', such as the creation of sperm from stem cells so that women can have children without male involvement in future. To add to their insecurity,[54] men are constantly reminded that the Y chromosome, which defines a male, is the smallest in the genome, suggesting it might have outlived its usefulness. Once upon a time, the Y chromosome was as large as the X chromosome. For some reason, about 166 million years ago, it began to lose the genes it contained. In humans, the Y chromosome now carries only 19 genes out of the 800 it originally shared with the X. Scientists predict that if the loss continues at this rate the Y chromosome will disappear altogether in about 4.6 million years. But men should take heart. Comparison of the genetic makeup of the Y chromosome in humans and chimpanzees shows a 30 per cent difference between the two primate species. In contrast, there is only about 2 per cent difference in the makeup of other chromosomes. What this means is that the Y chromosome is changing faster than other chromosomes, indicating it is not in evolutionary decline but is continuing to serve some essential function. One likely function is in reproduction. Size does not matter after all: small but perfectly formed will suffice.

Biologists have long been puzzled and continue to be puzzled by why genomic imprinting has evolved. What purpose does it serve? The aim of sexual reproduction, after all, is to harness genes from both parents in order to guard against any potentially defective genes that might be passed down from one parent alone. To confine certain activities to only one set of parental genes while switching off the other would seem to be counterproductive. The most widely held theory for the origin of genomic imprinting is that in every pregnancy there is conflict of interest between father and mother in the allocation of resources for the growth of the fetus. This has been described as the parental 'tug-o-war', or 'conflict'.[55] The father wants to maximize resources to benefit the offspring he has sired. That is why his genes are directed towards growth of the placenta which is the organ involved in the acquisition of these resources. The mother, on the other hand, would prefer to conserve resources to ensure her own survival and to share these resources among future offspring were she to breed again. The mother knows she will be related to all her offspring, present and future, whereas the father always has a residual uncertainty over whether any future babies will be his. The fact that genomic imprinting is seen only in placental mammals seems to support this view. Imprinted genes are not found in monotremes nor in fish, amphibians, and reptiles.[56]

As already noted in Chapter 1, many flowering plants develop a structure, the endosperm, which nourishes the embryonic seed in a manner analogous to the mammalian placenta. Like the placenta, the endosperm is also subjected to imprinting, with some genes active from the father and others from the mother. The gene that promotes the development of the endosperm in maize is the Maternal Expressed Gene 1 (*Meg1*), so in plants as in mammals, the mother takes charge of the allocation of nutrients to the embryo. For two such distinct structures, which evolved separately over millions of years, to adopt a similar mechanism like genomic imprinting clearly shows that this is the best way to balance paternal and maternal control in a situation where there is potential for parental conflict.

The battle of the sexes begins soon after the sperm fertilizes the egg, starting with the differing roles of paternal and maternal genes in the formation of the placenta. The battleground then moves to the implantation site in the uterus where the invasiveness of the fetal placenta is countered by the restraining influence of the maternal decidua as we have already seen in Chapter 2. There are, then, two possible areas of tug-o-war between father and mother acting at different times during pregnancy. One is between paternal and maternal genes in the formation of the placenta and the other is during the interaction between the invading fetal placenta and the resistance provided by the maternal uterine decidua. Since the superficial attachment of the epitheliochorial placenta is less intrusive to the mother than the haemochorial variety, the former presents less opportunity for fetal manipulation of maternal resources. This reduces the intensity of any confrontation that might arise. Different levels of paternal/maternal conflict could, therefore, explain the divergence in placental types rather than efficiency of nutrient transport. It is fortunate that this battle of the sexes in mammalian reproduction is relatively mild and innocuous compared to that seen in some insect species in which the female devours the male immediately after mating.

To portray parental genes as two warring factions may be misleading. Is genomic imprinting really a contest between father and mother to determine who wins? This kind of contest will benefit neither party because whoever triumphs will disrupt embryonic development. It must always be a draw, otherwise no pregnancy will result. Maybe genomic imprinting is not a battle after all but is parental cooperation rather than conflict. There must be an element of give and take. This seems to me to be a more rational interpretation. Throughout this book, we will encounter many examples of maternal cooperation rather than antagonism in reproduction.

As a general rule, paternal expressed imprinted genes promote growth, whereas maternal expressed ones tend to restrict growth. This is demonstrated very well by the first two imprinted genes identified in mice responsible for making a protein called Insulin Growth Factor 2 (IGF2) and for its corresponding receptor IGF2R.[57] A protein can only function when it is bound to its receptor, so they need to work together. These two genes are reciprocally imprinted, meaning that while the gene for *IGF2* is active from the father, the gene for its receptor *IGF2R* is the other way round, being active from the mother. IGF2 promotes growth of the placenta whereas IGF2R restricts this growth by mopping up excess IGF2 protein so there will be less of it available. In this way, maternal *IGF2R* genes counteract the growth effects of paternal *IGF2* genes in placental development. Both these genes are subjected to imprinting in eutherians and marsupials[58] but not in the egg-laying monotremes, showing that imprinting of these two genes only emerged after monotremes had branched off from placental mammals. In experiments with knock-out mice where the *IGF2R* gene is artificially disrupted, the offspring and placentas are much larger at birth than in normal mice and generally do not survive.[59] This observation confirms that maternal restraint via a functional IGF2R to counteract the growth promoting effects of the paternal IGF2 protein is essential for successful reproduction. Again, it comes down to a balance provided by opposing information from the father and mother.

A more sophisticated technique developed by Miguel Constancia has now made it possible to disrupt the *IGF2* gene only in the placenta while leaving the same gene in the fetus untouched.[60] This separation of the two targets is very useful because it allows us to see how placenta and fetus interact with each other. During early pregnancy, the placenta, deprived of the growth-promoting influence of IGF2, is small while the fetus remains normal size since its *IGF2* gene is not disrupted. However, by the later stages of gestation fetal growth also begins to slow. What is happening is that, at first, the placenta, even when small, manages to transport enough nutrients to the fetus but this becomes unsustainable later on. The placenta tries hard to compensate for its reduced size but can only do so for a while. Eventually, fetal demand for food will outstrip the placenta's ability to supply so the fetus suffers. This experiment is a very elegant demonstration of how fetal growth can be affected when placental function is compromised.

Too much IGF2 in humans can lead to a disorder called Beckwith–Wiedemann Syndrome (BWS).[61] Children with this condition are over-large and are prone to develop cancer early in life, such as Wilm's tumour of the kidney. The *IGF2* gene is paternal expressed. The reason for an increase in IGF2 is that the maternal *IGF2* gene, which should be silent, is accidentally switched on, resulting in both paternal and maternal copies being active. BWS is a good illustration of what can happen if there is disruption of the imprinting status of a single gene.

Sometimes, instead of involving a single gene, disruption can affect an entire chromosome. Chromosomes are long stretches of DNA along which genes are located. Humans have 46 chromosomes in the nucleus of each cell. Normally, chromosomes come in pairs, one from each parent. Occasionally, a mistake occurs, with a person inheriting two copies of the same chromosome from one parent and none from the other parent, a condition known as Uniparental Disomy (UPD). If both copies are from the father, it will be paternal UPD; if from the mother, it is maternal UPD. Many cases of UPD have no effects on health because most genes are not imprinted, so it does not matter whether a person inherits both copies from one parent instead of one copy from each parent. But it matters if the gene involved is an imprinted gene because the activity of such a gene is dependent on parent-of-origin.

Let us see what happens in two human diseases. The first disorders of UPD described in humans were Prader–Willi Syndrome (PWS) and Angelman Syndrome (AS). Both these conditions are the result of UPD of chromosome 15. This chromosome is rich in imprinted genes. Imprinted genes tend to occur in 'clusters'. They are not scattered indiscriminately all over the genome but usually group themselves together in one or two chromosomes. The other chromosome with many imprinted genes is chromosome 11. This is where the *IGF2* and *IGF2R* genes are located. In paternal UPD of chromosome 15, there will be no activity of any maternal expressed genes because both chromosomes have come from the father. This gives rise to PWS. Similarly, there will be no activity of any paternal expressed genes in maternal UPD of chromosome 15, resulting in AS. These two conditions present with very different abnormalities. PWS is characterized by uncontrolled eating and obesity, while AS shows mental retardation and impaired speech. These disorders emphasize the essential feature of imprinted genes: they can have a different action depending on whether they come from the father or mother. Non-imprinted genes do not behave like this. They have the same function regardless of parental origin. We have already seen in Chapter 3 how this parent-of-origin genetic inheritance can explain the different characteristics of mules and hinnies.

An interesting situation regarding imprinting is seen with the X chromosome. The X chromosome is one of the two sex chromosomes in humans, the other being the Y chromosome. Females have a pair of XX chromosomes (one inherited from the mother and one from the father) while males are XY (the X is inherited from the mother while the Y is from the father). In order to redress the balance with the male who has only one X chromosome, the extra X chromosome of the female is permanently silenced early in development. This is the 'X-inactivation' hypothesis put forward by the geneticist Mary Lyon in the early 1960s.[62] The purpose of this exercise is to prevent the female (who has two X chromosomes) from getting too much of whatever is produced by the X chromosome compared to the male (who has only one X), a mechanism known as 'dosage compensation'. As we saw earlier, too much IGF2 will cause Beck-with–Wiedemann Syndrome.

In the mammalian fetus, the X chromosome designated for inactivation is selected at random.[63] It can be the one derived from the mother or from the father. This seems a logical way to proceed since the purpose is just to balance any X-linked products between males and females. It doesn't matter which parental X chromosome is silenced. But this is not so in the placenta. Here, the selection is not random.[64] There is imprinting. It is the paternal X chromosome that is preferentially inactivated. The purpose here is no longer just dosage compensation, otherwise random inactivating would do. Why is imprinting required? Since the male placenta has only a single X chromosome inherited from the mother and a female placenta has the X chromosome from the father preferentially silenced, this could be a device to ensure that any X-linked placental products are only from the mother. There will be no 'foreign' signals coming from the paternal X chromosome that might trigger potential rejection of the placenta by the mother's immune system. Imprinting of the X chromosome in the placenta, therefore, could have an immunological function. Marsupials are an exception to this general pattern. Their X-inactivation preferentially affects the X chromosome from the father in both fetus and placenta. We do not yet have a satisfactory explanation why this should be so. It must have something to do with their different mode of reproduction compared to eutherians like ourselves. We might have to think again about X-inactivation in the human placenta because recent research shows that the pattern is a mosaic, like a patchwork quilt, with large areas where it is the paternal X and other areas where it is the maternal X that is inactivated, so X-inactivation in the human placenta is not uniform throughout. We will have to wait and see what this observation means.

Although designated a sex chromosome, it is a curious fact that the vast majority of genes on the X chromosome have nothing to do with sex. Take, for example, the gene that codes for an enzyme called glucose-6-phosphate dehydrogenase (G-6-PD) which is involved in carbohydrate metabolism. It is, therefore, a basic enzyme for common 'housekeeping' chores. Then, there is the gene that controls blood clotting. Mutation of this gene leads to the disease haemophilia in which patients suffer from uncontrolled bleeding. Another is the gene that allows us to see multiple colours whose defect causes colour blindness. All these genes are X-linked. This is why these diseases are more common in males, because men have only one copy of the X chromosome. If this copy goes wrong, then men are in trouble. Females, on the other hand, have two X chromosomes. If one X becomes defective, the other X will normally compensate, unless she is unlucky enough to have both her X chromosomes affected. Why are these stray genes found on the X chromosome if they are not responsible for sex determination? Maybe they are a legacy from the past when the X chromosome was just an ordinary autosome and then were not discarded when the status of the chromosome was elevated to that of a sex chromosome.

Imprinted genes do not function solely to promote or restrain placental growth. They can affect reproduction in other ways. Barry Keverne, the Cambridge neuroscientist, was the first to draw attention to the relationship between the placenta and the brain.[65] Several imprinted genes expressed in the placenta are also detected in the brain. One such gene is the *Paternal Expressed Gene 3* (*PEG3*) which is found in the placenta as well as in an area at the base of the brain known as the 'hypothalamus'. A major function of the hypothalamus is to control maternal nurturing behaviour after birth, such as lactation and bonding with the young. These activities of the hypothalamus are under the influence of hormones produced by the placenta (see Chapter 7). In this way, the placenta continues to manipulate the mother even after birth to serve the interest of the newborn. Female mice with artificially disrupted *PEG3* show impaired behaviour towards their offspring and fail to nurture them properly.[66]

Many of the young die due to this lack of care. Thus, paternal genes such as *PEG3* can affect the viability of the newborn. Imprinted genes, then, can influence different stages of pregnancy, both before birth in placental development as well as after birth in affecting the nurturing of the young via the hypothalamus. This postnatal maternal care includes suckling. When a mother suckles her young, she is prevented from becoming pregnant again; Nature has devised her own contraception. Delaying the onset of the next pregnancy, in turn, allows the mother more time to look after the present offspring. The entire series of events follow each other in a meaningful sequence.

When did the maternal hypothalamus learn how to respond to placental signals since the two only just met during pregnancy? This education probably took place in the previous generation when the mother herself was in the womb. Here, the placenta and fetal hypothalamus developed at the same time, both being influenced by the same paternal expressed gene *PEG3*. In mice, the fetal hypothalamus undergoes its major phase of growth at the same stage of pregnancy as when the hormone-producing cells of the placenta begin to increase, so the timing in development of the two organs does coincide. These are also the two structures that consume the most energy during pregnancy. Growing up together as it were, under the guidance of the same gene, has enabled the two structures to adapt to each other's needs. The experience gained during this early phase of development will be put to good use in the next generation when the mature hypothalamus in a pregnant mother responds to hormonal signals coming from her own fetal placenta. This will continue through subsequent generations. It is no coincidence that many of these genes, such as *PEG3*, are paternal expressed. The rate of propagation of a gene which confers 'benefits' is greatly enhanced if it is transmitted through the paternal line due to the relative ease by which males can spread their genes compared to females. It has been observed in mice that, over several generations of natural breeding, an 'advantageous' gene would have expanded much more when paternal expressed compared to maternal expressed.

In experiments in which pregnant mice are starved for 24 hours, the function of the imprinted gene *PEG3* is found to be disrupted in the placenta. What is surprising is that *PEG3* continues to be expressed by the hypothalamus of the fetus. The placenta and hypothalamus, therefore, react differently when confronted with adverse conditions within the uterus. Furthermore, during the period of starvation, placental cells start to devour themselves, a process known as 'autophagy' (self eating). The contents released by the dead placental cells are used to sustain the fetal hypothalamus which continues to grow and behave as if nothing has happened. The placenta offers part of itself in order to preserve fetal brain development, a noble example of self-sacrifice.[67]

Ramadan is a fasting month for Muslims. They fast during the day but are allowed to break their fast at sundown. Pregnant women are given special dispensation not to fast but most prefer to abide by the religious custom. Doctors in Saudi Arabia noticed that babies who were in the second or third trimester of pregnancy during Ramadan, were subsequently born with small placentas while the birth weight of the babies themselves was not affected. Is this another example of placental sacrifice to maintain fetal development when the mother is starved? This adds to the already long list of critical functions performed by this remarkable organ.

The mouse androgenome and gynogenome that led to the idea of genomic imprinting are laboratory creations. I will now describe two naturally occurring human examples that are their exact replicas. These human examples also strongly support the differing roles played by paternal and maternal genes in placental development. In the 13th century, Countess Margaret of Flanders was reported to have delivered a placenta made up of 365 fluid filled vesicles like a bunch of grapes. Each vesicle was mistaken for an

aborted embryo so she was thought to have brought forth 365 infants at one birth. 'Half of these were baptised John and the other
half Elizabeth, while the odd one out was judged to be a hermaph-
rodite and buried without baptism.' We now recognize this condition to be an abnormality of the placenta which pathologists have
classified as a 'complete hydatidiform mole', usually abbreviated to
just 'complete mole'. An embryo is usually absent. Because there is
no fetal circulation, the fluid absorbed by the placenta has nowhere
to go but to accumulate in the placenta, forming local collections of
fluid-filled vesicles which resemble a bunch of grapes (*hydatid* = sac
filled with fluid) (see Figure 9). Histological examination of a complete mole shows extensive overgrowth of placental tissue but the
most pertinent finding with respect to genomic imprinting is that
all the chromosomes carrying the genes are derived from the
father.[68] There are no maternal chromosomes. Poor embryonic development, overgrowth of placental tissue, and exclusive paternal
genetic contribution are features similar to those of androgenomes
created artificially in mice.

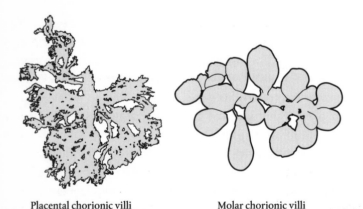

Placental chorionic villi   Molar chorionic villi

FIGURE 9 Comparison of normal and molar chorionic villi. Note the grape-
like vesicles in hydatidiform mole.

We have already seen earlier in this chapter what happens when there is disruption of imprinting affecting a single gene, such as *IGF2*. The result is Beckwith–Wiedemann Syndrome. Then there is disruption of an entire chromosome, such as UPD of chromosome 15, leading to Prader–Willi or Angelman Syndrome. Now we encounter a complete mole which represents a situation where there is disruption of imprinting involving the whole genome, that is *all* the genes.

How has this peculiar genetic makeup in a complete mole come about?[69] What does it tell us about imprinting? It is easier to explain this with the help of the following diagram, Figure 10. This might look complicated at first sight but, with a modicum of concentration, the reader will find it is easy to follow. It is worth devoting a few minutes to it. The effort will be worthwhile because the details are very revealing.

The first row of the diagram (1) depicts what happens normally when an egg is fertilized by a sperm. The resultant embryo will have one set of chromosomes inherited from the egg of the mother (M) and another set donated by the sperm of the father (P), making up the normal two sets (diploid) in the nucleus (MP).

The second row of the diagram (2) shows what happens in a complete mole. After fertilization of the egg by a sperm the maternal set of chromosomes (M) disappears. How this happens is not clear. Only the paternal set (P) from the fertilizing sperm remains. The paternal set of chromosomes doubles itself in order to restore the normal diploid complement of chromosomes in the nucleus. Without two sets of chromosomes to make up the correct number, a cell is not viable. If the fertilizing sperm has an X chromosome, then doubling up will produce an XX mole whereas a Y-bearing sperm will give rise to a YY mole. This latter is also non-viable which explains why complete moles have an XX composition. The reader will note that, although the final number of chromosomes attained is diploid, the chromosomes in both cases are exclusively from the

1) Normal fertilization

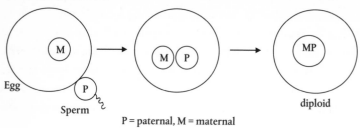

2) Complete hydatidiform mole

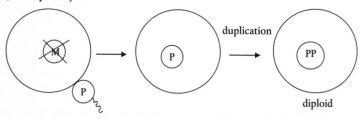

3) Partial hydatidiform mole

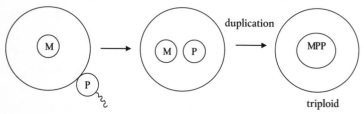

FIGURE 10 Mechanisms for the formation of complete and partial hydatidiform moles.

father (P). There is no maternal contribution (M). This is why complete moles are androgenetic. In 1967, I published a paper observing that nearly all complete moles had two X chromosomes and concluded that 'female' placentas, for some reason, were more prone to become moles.[70] Of course, I did not know then that both the X chromosomes in a mole were inherited from the father, unlike in a normal female in which one X is from the father and the other from the mother. Also, at that time, neither the concept of genomic imprinting nor the androgenetic nature of complete hydatidiform moles had been established. So I was correct in my observation but wrong in my interpretation. In science, as in other walks of life, appearance can be deceptive.

The third row of the diagram represents the formation of a partial mole (3). As its name implies, only part of the placenta is involved in the molar transformation, in contrast to a complete mole where the whole placenta is abnormal. The formation of a partial mole follows the same mechanism as for a complete mole but there is one crucial difference. In a partial mole, the maternal set of chromosomes in the egg does not disappear but is retained so that the resultant product contains three sets of chromosomes (triploid) rather than two (diploid): one set from the mother (M) and two sets from the father (P). The final outcome then is while *all* the chromosomes in a complete mole are derived from the father, the paternal contribution makes up only two thirds in a partial mole, with the remaining one-third being from the mother.

Taken together, these observations emphasize the fact that retention of maternal genes do provide some restraint in that only part of the placenta is involved in the molar transformation in partial mole rather than the whole placenta as in complete mole where only paternal genes are present. Maternal genes could provide something which prevents the formation of a mole. A protein designated as $p57^{KIP2}$ has been identified in normal placentas but is absent in complete moles. This protein is a product of a maternal gene known as *CDKNIC*. The lack of this maternally produced protein could trigger the formation of a mole, although how this works is not known.

A patient with a complete or partial mole would want to know whether she will have another with her next pregnancy.[71] The overall risk of this happening is about 2 per cent which is no higher than following a normal pregnancy. This statistic should allay to some extent any concern these women might have. But the risk rises to about 20 per cent after two or more complete moles.[72] Sporadic cases of women with 9–10 consecutive moles have been reported.[73] These recurrent molar pregnancies tend to congregate in families where there is a high degree of inbreeding, suggesting some form of recessive inheritance.[74] Interestingly, these familial recurrent moles do not show the exclusive presence of paternal genes as in sporadic moles. Instead, both paternal and maternal genes are present in equal numbers so they are known as 'biparental' moles. Their mechanism of production remains a mystery.

In Chapter 3, when describing choriocarcinoma, we said that the cancer can arise from a placenta following normal pregnancy. The risk is 1000 times higher after a complete mole. In the UK, approximately 1 in 12 patients with complete moles will develop choriocarcinoma. The risk for a partial mole is considerably lower. This suggests that the presence of paternal genes is likely to play an important role in the growth of choriocarcinoma, especially in the absence of maternal restraint.

A condition which is the exact opposite of a hydatidiform mole is seen in a tumour arising from the ovary known as 'teratoma'.[75] In this tumour all the genes are derived from the mother since it is from her own ovary. There are no paternal genes so it is analogous to the gynogenome created experimentally in mice. This difference in genetic constitution is reflected in the appearance of these two conditions. While a hydatidiform mole presents with an abundance of placental tissue, ovarian teratoma produces a variety of mature fetal tissues such as skin, hair, brain, muscle, teeth, and so on. How an ovarian teratoma is formed is not known. One possibility is that it is the result of fusion between two eggs in the ovary, a kind of parthenogenetic fertilization. Its attempt to form a proper embryo is thwarted by the absence of complementary paternal genes. It starts to make fetal tissues but these do not progress any further. Instead of a fully formed embryo the result is a haphazard collection of fetal tissues which collectively become a teratoma. These observations in hydatidiform mole and teratoma further confirm the importance of paternal and maternal genes in fetal and placental development. It is always encouraging and satisfying to have laboratory findings supported by naturally occurring examples.

We do not know why hydatidiform moles occur. These abnormalities are found only in humans. They are more prevalent in certain populations such as in Asia and some parts of South America. When the prevalence of a disease differs between countries, the finger of suspicion regarding the cause points to either some environmental factor present in the affected country, or that the population there has a greater inherent susceptibility to the disease. In a survey of the multiracial community in Hawaii, the incidence of hydatidiform moles was found to be considerably higher among those of Japanese or Chinese ancestry compared to the Caucasian or Hawaiian groups, suggesting that racial susceptibility may be more important than geographic location in the causation of these tumours. On the other hand the incidence of hydatidiform moles has fallen in recent years, suggesting that there might also be a large environmental component in their development. There is an increased risk of developing hydatidiform moles at the extremes of reproductive age. Girls under 15 years and women over 40 years of age are more at risk. None of these observations have been satisfactorily explained.

While the theory of genomic imprinting is currently a major focus of interest among developmental biologists, the mechanism by which imprinted genes are switched on or off is also attracting a lot of attention. This is because the way it is done represents a novel form of gene regulation. Up until now, genetic dogma has always dictated that to change the function of a gene its structure must be altered, which must happen by mutation. Biologists have now discovered that this is not always necessary. Gene expression can be controlled from the outside without disrupting the gene itself. Certain molecules, known as 'methyl groups', can attach themselves to the outside of the gene to switch off its function, a process called 'methylation'. Genes that continue to be active remain unmethylated. Since the mechanism does not involve alteration of the gene itself, it is designated as 'epigenetic'.[76] The gene has long been regarded as the master switch controlling all cellular functions. Now, we find there is another switch operating from outside the gene that can switch this master switch on or off. What this means is that gene function can be influenced by factors in the environment. Epigenetic mechanisms can override the primary genetic instructions so that some genes are selected to be used while others are not. Our genetic 'book of life' is merely a guide as to what is available. This has opened up a whole new field of research. The great advantage of epigenetic gene control is that it is faster and more flexible than mutation, allowing the cell to adapt quickly to cues from the environment.

Imprinted genes, as well as non-imprinted genes, can be affected by epigenetic modification. What effects does this have on the placenta? Imprinted genes are particularly sensitive to environmental influence because only one parental copy is normally active. There is no back up from the other parental gene as this one is silent. Since the placenta has a large number of imprinted genes, it is very susceptible to epigenetic modification. Its genes can be readily altered by the nutritional environment inside the womb, which, in turn, is a reflection of the state of the mother's health and fitness. The mother, then, makes a dual contribution to the development of the placenta. Like the father, she donates her genes when her egg is fertilized by his sperm but she also influences placental gene expression inside her uterus by epigenetic mechanisms. We shall return to this in more detail when we discuss 'fetal programming' in Chapter 5. So, from early development all the way to postpartum lactation and nurturing of the young, the mother shoulders a greater burden than the father. Parental involvement in reproduction is complementary but unequal.

Scientists are now becoming increasingly concerned that certain defects found in babies conceived by Assisted Reproductive Technology (ART) such as *in vitro* fertilization (IVF), might be caused by failure of proper genomic imprinting due to the artificial environment encountered by the fertilized egg during culture in the laboratory.[77] Starting life in a plastic dish in an incubator and immersed in an artificially made-up fluid can never be the same as an embryo implanted naturally in the womb, no matter how closely the laboratory conditions are tailored to mimic those found in the uterus. It is already well documented in animals that there is a high frequency of placental abnormalities and reduced survival rate among offspring conceived by ART. Similar findings are now seen in humans where a two-fold increase in the incidence of congenital malformations has been reported. Diseases like BWS described earlier occur in about 1 in 15,000 natural births, but in babies born from ART the incidence rises to as high as 1 in 4000. The placentas of babies conceived by IVF have low levels of 'methylation' of DNA.[78] We noted earlier that this is the epigenetic mechanism by which imprinted genes are switched off. What this implies is that in the placentas of IVF babies, genes which should be switched off during imprinting are not and they remain active because of the lack of methylation. Many types of cancer also show low levels of methylation, suggesting that their epigenetic control mechanism could be dysfunctional, allowing genes which should be silent to become active again. This might account for the erratic and unrestrained behaviour of cancer cells.

Robert Edwards and Patrick Steptoe, the pioneers of IVF, described their technique in 1969 in a paper published by the scientific journal *Nature* and the world's first 'test tube' baby, Louise Brown, was born in Oldham Hospital in July 1978. Since then, in many developed Western countries, between 2–4 per cent of all births are conceived by ART such as IVF. This is an enormous number and is likely to rise further as ART is now a booming industry which is not going to be reversed. In animal breeding programmes, IVF is beginning to achieve a very high success rate with improving technology. Soon this will be the case for humans. When this happens then IVF is likely to be the preferred choice for infertile couples rather than embarking on the vagaries of 'trying for a baby' the natural way. If these IVF babies carry a high risk of abnormalities then the long-term future of human reproduction could be severely jeopardized. Furthermore, epigenetic changes are known to be transmissible to subsequent generations so defects in IVF babies could affect their own children. All this is rather worrying.

Disruption of imprinting is particularly relevant to reproductive 'cloning' although at present this procedure is performed only in animals. In 1997, Dolly the sheep was the first animal produced by cloning using a technique known as Somatic Cell Nuclear Transplant (SCNT). This involves fusing a normal adult cell from a sheep (A) with the egg of another sheep (B) from which the whole nucleus (and therefore all the genes) has been removed. The fused egg is then artificially stimulated to develop into a full embryo. This embryo of course will contain only the genes from sheep (A) while the enucleated egg from sheep (B) is used merely as an incubator. The genes therefore are all from one animal (A). This is the main purpose of cloning: to reproduce a progeny identical to the donor. It will be immediately apparent that the normal process of genomic imprinting will not occur since all the genes are from one animal. There are no paternal and maternal contributions as in normal pregnancy. The success rate of cloning is extremely low with the major defects being in the placenta. It took 277 attempts before Dolly was born. Since Dolly the sheep, reproductive cloning using SCNT has been tried in other animals such as cattle. Again, pregnancy failure is very high with most losses due to abnormal placentation. Those that survive to be born usually die soon afterwards. A few succeed in reaching adulthood like Dolly, but they do not live long due to failure of various systems inside the body. Dolly, herself, died at the age of six, which was half the typical lifespan of the species of sheep to which she belonged. The vision of some mad scientist at a secret hideout manufacturing human clones to be released into an unsuspecting world must remain in the realm of science fiction—at least for now.

5

Connecting with Mother

At first the blastocyst lying inside the uterus feeds off the secretions from the uterine glands (uterine milk).[79] This is adequate for the first few weeks of pregnancy, but soon the rapidly growing embryo needs to tap into a more abundant source of food supply. For this, it has to develop a placenta. The first step in making a placenta is for the blastocyst to attach itself to the lining of the uterus and then burrow into the underlying tissue, a process known as 'implantation'. This is the most crucial stage along the production line towards making a baby. If pregnancy fails, this is where it will usually occur. Many cases of early miscarriage are the result of deficient implantation. In humans, implantation begins 6–7 days after fertilization of the egg. In some animals, implantation can be actively delayed in response to seasonal cues or other environmental changes that might make the world outside the uterus more hostile. By delaying the birth of the baby until the time is right, this device optimizes the chances of survival of the offspring. Kangaroos do this all the time but humans have either lost or never acquired this rather useful skill.

Implantation begins when the trophoblast layer surrounding the blastocyst makes contact and sticks to the surface epithelial lining of the uterus.[80] Before this can happen the blastocyst has to emerge from a thin membrane called the 'zona pellucida' which encases the blastocyst like an eggshell. Not surprisingly this phase is known as the 'hatching' phase. The trophectoderm is now free to make contact with the lining of the uterus. Microdissection experiments of mouse blastocyst into its separate components of the inner cell mass (ICM) and trophectoderm (see Chapter 3) show that only trophectoderm will implant when transferred into a uterus. The ultimate fates and behaviour of the two cell lineages are already decided at this very early stage of development.

The site of implantation determines the final position of the placenta. This is usually on the back or side walls of the body of the uterus but can sometimes occur in inconvenient places such as over the opening of the cervix, resulting in a condition called 'placenta praevia' which will cause obstruction during the birth of the baby. As we saw in Chapter 2, when the placenta overlies an area of the uterus deficient in decidua, such as an old Caesarean section scar, there is a danger the placenta may adhere too firmly and may not separate from the uterus after birth, a condition known as 'placenta creta'. Both these conditions require surgical intervention. Implantation can also occur in the Fallopian tubes, leading to tubal or ectopic pregnancy with its associated complications such as rupture of the tubes, which is another obstetric emergency. Pathologists have argued about what causes this rupture. Some believe it is purely mechanical due to the mismatch between the increasing size of the growing embryo and the small, unyielding lumen of the Fallopian tube. Others think this view is too simplistic and propose that absence of the restraining influence provided by decidua is to blame. Histological examination of the implantation site in tubal pregnancy does show a very different picture to that seen in uterine pregnancy. In the tubes, there are few signs of decidual formation and trophoblast invasion appears to be uncontrolled, penetrating deep through the mucosal lining and muscle layer to reach as far as the outermost peritoneal surface. With this kind of aggressive behaviour, eventual rupture should come as no surprise.

A great deal of hype was generated in 2007 when the media showed a picture of a heavily pregnant man complete with facial hair. The main reaction was one of disbelief. Was it a hoax? Was it really possible? The answer was 'yes', it was possible. In this particular case Thomas Beatie, as he is now called, was originally a woman named Tracy LaGondino, before she underwent transgender conversion into a man. But residues of female organs such as the uterus were left behind, so an appropriate receptacle for the implanting placenta was already in place. Even in the absence of a uterus, the placenta can implant into other organs. Placental trophoblast cells are inherently programmed to be invasive no matter where they are. Hertig likened the implanting blastocyst to a hand grenade. Once the pin is pulled there is no way to stop it exploding. As we have already seen, implantation can take place in the Fallopian tubes. Occasionally it can also occur over the intestines, creating what is known as 'abdominal pregnancy'. Theoretically any organ (in a man or a woman) will do if it has the necessary blood supply that can be tapped by the placenta. But these extra-uterine sites do not have the restraining influence of decidua which is produced only by the uterus so that trophoblast invasion will be totally uncontrolled. The ultimate danger, then, will be catastrophic bleeding in an individual pregnant without a uterus.

There is much excitement over claims that artificial wombs might, one day, be created in the laboratory. This would indeed be a major advance in reproductive technology if and when it happens. The womb is not just a receptacle for the baby. It is a highly sophisticated organ under hormonal control which would be extremely difficult to copy. Transplantation of a womb from another individual might be more feasible but even here the problems are not trivial. The technical difficulties (mainly related to providing an adequate blood supply to the transplanted womb) are said to have largely been solved, at least in experiments with rabbits, although many of the animals did not survive. (This reminds me of the aphorism every medical student remembers: 'the operation was successful but the patient died'.) The experiments, therefore, should be considered only partially successful. Nevertheless, human clinical trials might start soon. There is even the possibility of a recipient receiving a womb transplant donated from her own mother. The patient then will house an organ in which she herself had spent some time in early life, an exemplary use of recycling. There are many ethical questions that need to be answered. Organ transplantation is performed as a life-saving measure by replacing a failing vital organ with a new one. Otherwise the patient will die. The risks involved are justified. The same cannot be said for a healthy woman wanting a new womb in order to conceive. Failure to have children is not a life or death issue. To subject a healthy woman to such a risky procedure is ethically questionable. Also if anything were to go wrong with the procedure it would affect the lives of three individuals—the donor of the womb, the recipient, and the baby in the transplanted womb, assuming that implantation was successful. However, womb transplants do have one advantage over conventional organ transplants in that the former do not have to last the lifetime of the recipient. They only have to function long enough in the recipient to produce the number of children required. After that the transplanted womb can be removed. The recipient will not be subjected to a lifelong regime of immunosuppressive drugs with all their unwanted side effects. And if a womb transplant can be offered to a woman, why not also to a man?

The surface of the uterine epithelial lining does not normally permit attachment by the blastocyst and only becomes receptive in a specific period under the stimulus of hormones.[81] This period of receptivity known as the 'implantation window' is estimated to last about 6–8 days after fertilization.[82] Outside this window implantation does not occur. We still do not know for certain what happens to the uterine epithelium to make it more receptive. Changes in its physical and biochemical properties induced by pregnancy hormones are likely but what these changes are remain to be defined. Paradoxically, there is no resistance to implantation outside the uterus. A blastocyst can successfully adhere to any organ in the abdomen at any time without the need to coincide with a period of receptivity. This results in abdominal pregnancy. The long periods of uterine epithelial hostility to the blastocyst interspersed with short spans of acceptance would allow the uterus a degree of control over the best time for implantation. The mother makes the decision, not the embryo. Nature has not seen the need to provide this regulatory device to extra-uterine sites that are not normally involved in implantation.

The stage of development of the embryo must also coincide with this 'implantation window'. This is obviously an important area of research because the small number of successful pregnancies achieved by in vitro fertilization (IVF) relative to the high rate of successful fertilization is mainly due to failure to synchronize the state of embryonic development in the laboratory with this 'implantation window' in the uterus. Only about 30 per cent of embryos transferred after IVF lead to the birth of a healthy baby, depending on the centre involved and the maternal age. There is a huge wastage which is very likely due to failure to implant. To improve the rate of success, more than one embryo per transfer is generally used.[83] This has resulted in an increase in multiple pregnancies with the associated risks for both mothers and babies. For this reason, the Human Fertilisation and Embryology Authority (HFEA) in the UK strongly discourages this practice. The major hurdle in IVF then is not so much the fertilization of the egg with sperm, which can readily be accomplished, but the risk that when the fertilized egg is put back into the uterus it does not implant so that the overall success rate of 'take home' babies remains low. Although the early embryo can be nurtured for a brief period in the laboratory, it is not possible to rear it to term without implanting it in the uterus for further development. The term 'test-tube' baby is really a misnomer in that only the process of fertilization and initial stages of embryonic development take place in a test tube (*in vitro*), the subsequent growth has to be within the uterus. The idea of creating humans by ectogenesis (gestation outside the uterus) has been perpetuated by tales such as Goethe's *Faust* brewing a tiny human inside a flask or Aldous Huxley's *Brave New World* with its sinister hatchery replete with bottles of embryos. Attempts to make people by artificial means have a long and interesting history, as can be seen in Philip Ball's book *Unnatural—The Heretical Idea of Making People*.

In Chapter 3, we defined three populations of trophoblast which have arisen from the trophectoderm layer of the blastocyst. They are syncytiotrophoblast, endovascular trophoblast, and interstitial trophoblast. Each of these makes its own distinctive contribution to implantation. Syncytiotrophoblast is the first to penetrate the surface epithelial lining of the uterus and move into the underlying uterine tissue or decidua.[84] It does this by insinuating itself between the cells of the uterine epithelium and then squeezing through. When it encounters the uterine blood vessels, syncytiotrophoblast is able to erode and break down the walls of these vessels so that blood leaks out to form pools surrounded by syncytiotrophoblast, creating what is known as the intervillous space which is the first step towards the establishment of the haemochorial placenta.[85] The placenta has now become a parasite upon the mother. It has literally burrowed into the substance of her womb and is siphoning off nutrients from her blood to provide for the embryo. The erosion of uterine blood vessels by syncytiotrophoblast is a remarkable achievement because normal cells do not have the ability to break down blood vessels. Only cancer cells can do this, which is why cancer is able to spread (metastasize) into the patient's bloodstream. Cancer cells secrete an enzyme, collagenase, that disrupts the walls of blood vessels. It is likely that syncytiotrophoblast also produces a similar enzyme.

While the syncytiotrophoblast is completing the initial stages of implantation (see Figure 11), the underlying endovascular and interstitial trophoblast will also start to invade the uterus. Now it is the turn for these other two trophoblast populations to be useful. They have a very important job to do. They break through the overlying layer of syncytiotrophoblast and move down into the decidua underneath. Endovascular trophoblast advances down the openings of the uterine arteries to form loose clumps which plug the lumen within the arteries. This is why they are called endovascular trophoblast because they lie inside the uterine blood vessels. It is only

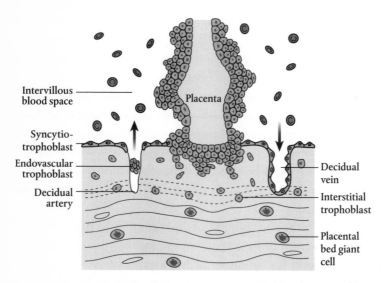

FIGURE 11 Diagrammatic representation of the implantation site showing the locations of the different trophoblast populations. The two arrows indicate the directions of blood flow to and from the intervillous space.

recently that we have understood why these plugs form.[86] They have two functions. One is to prevent maternal blood from the uterine arteries reaching the developing embryo at too high a pressure with the attendant danger of 'blowing out' the embryo. The other is to restrict the amount of oxygen reaching the embryo. This may seem paradoxical since a major function of the placenta is to supply oxygen to the embryo. Indeed, a threshold level of oxygen is necessary, since few existing mammals are capable of reproducing above an altitude of 4500m where there is not enough oxygen. However, the early formation of embryonic organs requires an environment with a low oxygen content. Otherwise the generation of too many high reactive oxygen free radicals will damage the cells. Embryonic cells, in particular, are highly susceptible to what is termed oxidative stress because they do not have enough anti-oxidant enzymes to neutralize the oxygen free radicals. The benefits of oxygen come at a price. Too little is insufficient to generate enough energy while too much is damaging. Graham Burton, who made the original observation of the endovascular trophoblast plugs with Jans Brosen, refers to oxygen as the 'Janus' gas, from the Roman god Janus who is depicted with two heads facing in opposite directions. So, although the ultimate aim of implantation is for trophoblast to make connection with maternal blood to establish the haemochorial placenta, this has to be tempered, at least during the first few weeks, by a mechanism by which this blood supply with its oxygen content is appropriately controlled. The trophoblast plugs provide this control.

It is interesting that early development of the human embryo should utilize a metabolic pathway low in oxygen similar to that employed by ancient species many millions of years ago when oxygen had not yet become plentiful on Earth.[87] There is a grandiose sounding phrase to describe such a phenomenon, coined by the German evolutionary biologist Ernst Haeckel: 'ontogeny recapitulates phylogeny'. What this means simply is that during human embryonic development (ontogeny) there are stages which resemble the features of a more ancient species from which we have descended (phylogeny). We display our evolutionary history as an embryo. While this concept is now considered to be too broad a generalization, some examples do exist. For example, very early human embryos have slit-like structures similar to the gills of fishes, reminding us that our ancestors once lived in the sea. Tom, the chimney sweep boy in Charles Kingsley's *Water Babies*, discovered he had sprouted a set of external gills when he woke up in his magical underwater world. He had recaptured an ancient trait.

Eventually these endovascular trophoblast plugs will break up and maternal blood can then flow freely towards the embryo at a stage when it is more firmly anchored and no longer as vulnerable to high levels of oxygen. Once the plugs are broken up, some of the trophoblast cells which are released start to spread down the inside of the arteries along the inner wall, elegantly described by Elizabeth Ramsey as 'like wax dripping down a candle'. This migration takes place against the direction of blood flow, so they are not just passively swept along by the current. It must be an active process, implying that there is a purpose to the exercise. During the migration some of these trophoblast cells become incorporated into the substance of the wall of the arteries thereby altering their structure. This is endovascular trophoblast's second function.

Meanwhile, interstitial trophoblast cells invade into the substance of the decidua outside the arteries.[88] They do not proceed any further once they have gone through the decidua and have reached the muscle layer of the uterus (myometrium). This depth of invasion is always the same in every pregnancy indicating that it is not random. Some very fine tuning is involved. At this juncture, large numbers of multinucleate cells—known as placental bed 'giant cells'—are present. These cells represent the final stage of trophoblast differentiation. They signal that nothing more will happen; it is the end of the line. What is the signal that stops the migration? Who gives it? Is it predetermined in the placenta or is it limited by the uterus? These questions will be discussed in subsequent chapters. Occasionally, this control breaks down, allowing trophoblast to penetrate the uterus deeper than it should. Depending on the depth of trophoblast invasion, the clinical conditions resulting from this lack of control are placental 'accreta' (when trophoblast reaches beyond the decidua), 'increta' (trophoblast found deep in the muscle coat), and 'percreta' (trophoblast penetrates right through and perforates the uterus). All these are potentially dangerous complications.

In the decidua, interstitial trophoblast cells tend to move preferentially towards the decidual arteries to surround the outside of the blood vessels. They have a remarkable effect on these blood vessels whose walls begin to show signs of damage and degeneration. These surrounding interstitial trophoblast cells secrete some kind of enzyme that digests the walls of the arteries from the outside. The overall effect of these two routes of trophoblast migration is to alter the structure of the walls of the decidual arteries, acting either from the inside by endovascular trophoblast or from the outside by interstitial trophoblast. There is still some debate as to which occurred first. Experts in this area of research, like Robert Pijnenborg, believe that destruction of the arterial wall initially comes from the outside. Once this 'priming' effect has taken place, the plugs of trophoblast cells which have spread down the lumen become embedded in the wall from the inside.[89]

No matter what the sequence of events, this destruction of the walls of the decidual arteries by trophoblast (also referred to as 're-modelling') is crucial to successful pregnancy. Its importance cannot be over-emphasized because the process enables blood to be delivered at the right speed and volume to the growing embryo and the expanding placenta. Before remodelling, the decidual arteries are small calibre vessels with muscular walls which are subject to nervous control like other arteries in the body. Blood is normally shunted around the body according to where it is most needed and this is regulated by the nervous system contracting or relaxing the muscles of the vessel wall. Once the muscles of the arterial walls are destroyed, the decidual blood vessels are liberated from this controlling mechanism so that blood cannot be diverted away from the placenta even if the mother engages in strenuous exercise. This ensures there is always enough blood flowing to the fetus and placenta regardless of what activities the mother is engaged in. A second very important consequence of this arterial remodelling is that the once small calibre uterine arteries are now converted into larger diameter vessels (see Figure 12). Harries and Ramsey, in a painstaking 3-dimensional reconstruction of serial histological sections from pregnant uteri, showed a five-fold increase in diameter especially at the mouths of these arteries before they open into the placenta, like the expanded end of a funnel. This will have a profound effect on the blood flow into the placenta. It converts the normally high pressure, high velocity uterine arterial system into a low pressure, high volume blood flow. This means that blood from the uterus reaches the placenta in adequate amounts and at sufficient momentum to maintain blood flow inside the placenta but is not so strong as to cause damage to the placenta.[90]

Once the uterine arteries have been transformed by trophoblast during the first pregnancy, they never return to their original shape. There is no complete repair to the normal smooth muscle wall of the arteries and the elastic tissue there remains fragmented. This might explain why second and subsequent pregnancies are less prone to complications than the first because the residual effects of arterial remodelling left over from the first pregnancy are still present. The reason why uterine arteries have to undergo this complex process of remodelling every time pregnancy occurs is that they have conflicting roles to play. In the non-pregnant state they need to be small calibre vessels to restrict blood loss during menstruation, whereas in pregnancy they have to behave like dilated vessels to allow adequate blood flow to the fetus and placenta.

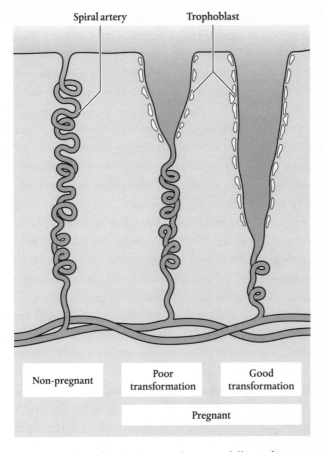

FIGURE 12 Complete and incomplete remodelling of uterine spiral artery by trophoblast.

In normal pregnancy, the calibre of the uterine arteries can be increased to a radius of as much as 250μm, but this increase is usually not more than 100μm when there is inadequate transformation. This can lead to a significant resistance to blood flow to the placenta, and will have very serious consequences. Depending on its severity, the baby might die early in pregnancy, resulting in miscarriage. Or

111

it might succumb later, presenting as still birth.[91] All this is extremely distressing to the parents especially when the cause is not obvious to them. With luck, the baby may survive but because of the lack of nutrients it will be born smaller than normal, a condition called intrauterine growth restriction (IUGR).[92] As we shall see later, these small babies are not healthy when they grow up. The final complication that can follow inadequate remodelling is the disease pre-eclampsia (PE). The cause of the first three conditions is clear: inadequate blood flow will disrupt the development of the embryo with varying degrees of severity. The events leading up to pre-eclampsia require further explanation.

Pre-eclampsia (PE) differs from the other conditions in that it affects both baby and mother.[93] The starting point is the same as for the others. Inadequate transformation of the uterine arteries by trophoblast will lead to not enough blood getting to the placenta which will start to degenerate. As a result, small fragments of syncytiotrophoblast will break off into the maternal circulation where they liberate a variety of substances that have harmful effects on the mother's blood vessels.[94] Some women are more susceptible to these substances than others. This is the ultimate cause of the symptoms of PE, the most conspicuous being a dramatic rise in blood pressure. Two sequential events are therefore involved in PE. The first is poor arterial remodelling resulting in inadequate blood perfusing the placenta. The injured placenta releases substances into the maternal circulation which are harmful to her blood vessels, but what these substances are has not been identified. It is this second event that leads to the symptoms of PE. This two-stage scenario adds to the complexity of the disease. Susceptibility to PE also has a large genetic component, for example women from Africa have a higher risk of developing PE, but at what stage do genes come into play? Are some women inherently more susceptible to the harmful debris released by a damaged placenta? In other women, does the fault lie in their being more resistant to having their uterine arteries remodelled by trophoblast? There is a great deal more to be learnt before we can fully understand the mechanisms underlying PE but we can be certain of one conclusion: trophoblast cells are involved.

There are two curious features about PE. One is that it tends to mainly affect women in their first pregnancy. As we noted earlier, once the uterine arteries are remodelled by trophoblast they do not return entirely to their original state, so that blood flow to the placenta is easier to sustain in the second and subsequent pregnancies. This could explain the lower incidence of PE. The second feature is that PE occurs less frequently in women who smoke. This is perhaps the only disease in which smoking seems to be beneficial. There is presently no obvious explanation. Something in a smoker's bloodstream might protect against the harmful effects of the substances released by trophoblast.

If untreated, PE is followed by convulsions and death of the mother. PE is a serious obstetrical complication in all parts of the world. About 1 in 2000 pregnancies are affected and it is the most common cause of maternal mortality. It has been known for almost 100 years that the disease originates in the placenta. The presence of a placenta is both necessary and sufficient to cause the disease. A fetus is not required because PE can arise in patients with hydatidiform moles where there is no fetus. An increase in placental mass such as in twin pregnancies is also associated with a higher frequency of PE. Even a uterus is not essential since PE is seen in abdominal pregnancies. The management of the disease then, is to deliver the baby immediately and remove the placenta. This has to be done whether or not the baby is old enough to survive. Otherwise both the mother and baby will succumb. How can a disease which predisposes to such a lethal condition in women of reproductive age continue to be present in populations worldwide? Why have the genes responsible for the condition not died out? Again, we come back to the idea of the tug-o-war between the placenta and the mother (see Chapter 4). Since the initial defect in PE is inadequate trophoblast invasion and remodelling of the uterine arteries, the genes which inadvertently lead to PE could be those maternal defence genes which are there normally to control over-enthusiastic trophoblast invasion. These must be important genes to conserve. If the balance is sometimes tilted too much in favour of the mother's defence, leading to adverse side effects like PE, this is the price that has to be paid.

The symptoms of PE can come on very suddenly. For this reason, a great deal of research is now focused on identifying potential markers that could predict its onset.[95] Several approaches are being tried, based on what we already know about the mechanisms involved in the production of PE. The first of these is to scan pregnant women by Doppler ultrasound, a technique that will show up any impedance of blood flow within the placenta since inadequate trophoblast conversion of uterine arteries into large calibre blood vessels is a fundamental defect in PE, as we have already described. Measuring the amounts of placental DNA in the mother's blood has also been tried. The initial event leading up to the development of PE is damage to the placenta due to insufficient blood supply. Dead and dying placental cells will shed their nuclear DNA material into the maternal circulation so increased levels of this DNA is a bad sign. The corollary to this is that a healthy placenta manufactures a wide variety of products essential for pregnancy (see Chapter 7). Decreased levels of these products in maternal blood would then signify a degenerating placenta, another potential predictive marker for PE. A combination of all these approaches is now being evaluated but, so far, none have yet been adopted as routine procedures in obstetric practice. This is because a crucial question remains. What can we do to prevent a woman from developing PE, even if we were to find signs that she is at high risk of doing so, except to monitor her closely? The search for predictive markers will be a wasted effort unless it can be followed up by successful preventive measures. These latter are not yet available.[96]

PE is a good example of where the lack of an animal model is a great disadvantage in attempts to understand the underlying cause of the disease. Animals do not suffer from PE nor do they have a similar pattern or depth of trophoblast invasion and artery remodelling as seen in humans. Even closely related primates are different. For example, in Old World monkeys such as baboons and Rhesus monkeys, trophoblast migration occurs mainly within the lumen of the uterine blood vessels with very little invasion through the substance of the decidua outside the blood vessels, that is by endovascular trophoblast and not by interstitial trophoblast. This pattern contrasts with that in humans where both routes of invasion are prominent. The pattern of trophoblast invasion in the Great Apes is more akin to humans. Pre-eclampsia has been described in gorillas and chimpanzees but the ethical issues relating to research on these species is probably just as problematic as for humans.

Miscarriage, still birth, IUGR, and PE are all visible legacies of poor trophoblast remodelling of uterine arteries. But there are other complications that are hidden from view and are not immediately apparent. Babies with low birth weight resulting from IUGR, although fortunate to be born alive, have not escaped unscathed. They are more prone to develop heart disease when they reach adult life. This association was reported by the epidemiologist David Barker in the late 1980s.[97] He sorted through the records of births occurring during and after the First World War and noticed a greater proportion of adults who started life as small babies subsequently died from heart disease. The National Institute of Child Health has now issued the following manifesto: 'Coronary heart disease, the number one cause of death among adult men and women, is more closely related to low birth weight than to known behavioural risk factors'. This points to IUGR being more important than previously recognized causes, such as smoking and lack of exercise. Barker related this to poor maternal diet during pregnancy leading to insufficient nutrients available for proper fetal development.

The heart is particularly vulnerable.[98] The construction of a muscular pump that will continue to beat uninterrupted for many decades must be a highly complex process. It is no wonder that even small mistakes while putting it together can lead to it malfunctioning later on. Formation of the heart can also be affected indirectly. A gene controlling normal blood vessel development in the placenta is HOXA 13. When the function of this gene is artificially disrupted in mice, blood vessels in the placenta do not develop properly, although the trophoblast layers remain normal. These mice die of cardiac failure. Examination of the heart shows a dramatic reduction in thickness of the ventricular walls which is why it cannot pump blood round the circulation. The interesting thing about HOXA 13 is that the gene is not active in the embryonic heart itself and yet the primary lethal defect is in this organ. What happens is that disruption of blood flow through the placenta places an unsustainable burden on the embryonic heart, a good example of how placental abnormality affects fetal heart development. Besides heart disease, small babies are also linked to other adult illnesses, including diabetes, cancer, and neurological disorders. It seems that nutrient deprivation in the womb at a critical stage of development can result in permanent damage to many of the baby's organs. Much of the placental growth takes place during the first half of pregnancy while the fetal growth curve shoots up in the second half. Whether the placenta or fetus will suffer most will depend on which stage food is scarce.

Of course, what food the fetus gets in the womb is not simply what the mother eats. It also relies on the placenta's ability to transfer this food across to the fetus.[99] This is why inadequate trophoblast transformation of uterine arteries and the resultant insufficient blood flow to the fetus is such an important cause of IUGR. The placenta can also control the flow of nutrients by altering the number and efficiency of transporters on its surface according to availability in the mother and fetal demand.[100] This will be discussed further in Chapter 8.

So, health in later life is strongly influenced by events in the womb.[101] Samuel Taylor Coleridge showed great foresight in penning the quotation at the beginning of this book. The traditional view of mammalian development is that it is pre-determined by the genes inherited by the embryo. The course is already set right from the start. But we now know that this is not necessarily so. The form and function of the embryo and its organs—its phenotype—can be modified by the environment inside the womb by epigenetic mechanisms which we already encountered in Chapter 4.[102] Embryonic organs have a degree of 'plasticity', meaning they can be moulded in different ways before settling down into their final form. This process is known as 'fetal programming'. Mammalian development is shaped by a combination of Nature via genetic inheritance and Nurture through environmental influence.

The distinction between Nature and Nurture has become blurred. The terms were coined by the geneticist Francis Galton, who was Charles Darwin's half cousin, over 150 years ago. Galton himself believed that Nature was more influential than Nurture in human development. With our increasing understanding of epigenetics and fetal programming, he might want to change his mind were he alive today.

A fetus, deprived of food in the womb, will make appropriate adaptations in its growth pattern and metabolism to take into account this poor nutritional environment. At birth, it finds its food intake and energy consumption is not in tune with the postnatal surroundings where everything is freely available, but it is now too late to change. This maladjustment between what the fetus is programmed to do inside the womb and the reality of the world outside leads to an increased risk of heart and other metabolic diseases like diabetes. Childhood and adult obesity might also start in the same way. An undernourished fetus *in utero* will be programmed to store more fat. It then finds this provisional store is no longer needed after birth but the fat remains. All this is detailed by Peter

Gluckman and Mark Hanson in their book appropriately entitled *Mismatch*.[103]

Boys are more affected by fetal programming than girls because they grow faster in the womb and are, thus, more susceptible to nutrient deprivation.[104] Boys are more prone to IUGR and, at times of famine, more boys succumb before birth. Immediately after birth, male neonates are also more likely to die. The compensatory mechanism for a poor intrauterine environment appears to be less efficient in the male placenta compared to the female. We have already alluded to this sad but incontrovertible fact many times in preceding chapters, that life is more hazardous for the male of the species. The sex of the placenta has never been fully taken into account when considering its function but this can no longer be ignored. The pattern of expression of many genes is now shown to be distinct in the male and female placenta. The placenta is not asexual and will behave differently in the two sexes.

Animal studies confirm the phenomenon of fetal programming. Minor alterations in the diet of pregnant animals are sufficient to produce lasting changes in the offspring. Even insects show examples of how diet influences fetal development. Biologists have long been puzzled by the marked difference between the queen bee and the workers in a honey bee colony in spite of being genetically identical. The queen is larger, develops faster, has a longer life span, and is reproductively more efficient, producing up to 2000 eggs per day. Note it is the *queen* and not the *king* bee who rules. Once a year, the queen mates with a male after which he is killed. It is a matriarchal society with the familiar theme of maternal control of reproduction so often encountered in mammals.

What makes the queen so different? It all stems from diet in early life. At the larval stage, the future queen is raised on royal jelly, a rich source of fats and proteins, while the worker bees make do with a less lavish meal. The difference in food source has a dramatic impact on development, especially of the brain, and determines who behaves like royalty and who becomes a humble worker. The active ingredient in royal jelly has been identified and is given the name of 'royalactin'. So royal jelly is not just a nutritious brew but contains a specific substance which is capable of transforming one type of bee into another. Surprisingly, royalactin has a similar action when given to the fruit fly *Drosophila*, transforming those larvae fed on it into queen-like individuals, even though fruit flies do not have queens in their social structure. Honey bees and fruit flies belong to different orders of insects that split off from each other many millions of years ago, and yet royalactin continues to function in both. Royalactin has obviously been put to good use by the honey bee in fetal programming but why does it continue to be active in fruit flies where it is not needed? This is a remarkable example of evolutionary conservation of a molecule. Royal jelly has been a popular human dietary supplement for a long time, especially in Asian countries. Maybe its imagined beneficial effects do have a scientific basis after all. If it works on fruit flies, why not with humans?

We have so far looked at the overall picture of the implantation site and have considered some of the consequences that might arise if trophoblast migration into the uterus is inadequate. Let us now examine more closely the mechanisms involved in this very delicate operation. The process begins with trophoblast interacting with some components within the decidua, as this is the layer through which migrating trophoblast will traverse. What might these components be? Cells generally require some kind of traction to move, something to hold onto. This is provided by 'cell adhesion molecules' (CAM).[105] As their name implies these are specialized molecules on the surface of cells which are used to anchor them to the glue-like substance surrounding each cell, like cement holding bricks together in a wall. In this way, the structural integrity of the tissue is maintained. The material providing the in-fill between cells is the extracellular matrix (ECM). For implantation, the important CAM on the surface of trophoblast cells belong to a family called 'integrins'.

These integrins will stick to two major classes of ECM produced by decidua, namely laminin (LM) and fibronectin (FN). The amounts of LM and FN in the uterus vary throughout the menstrual cycle, with very little present at the early phase of the cycle, then rising steadily to peak at the late stage of the cycle. If pregnancy occurs large quantities of LM and FN continue to be produced by cells of the decidua. This cyclical variation suggests that synthesis of LM and FN is under hormonal control and that they have an important role to play in the pregnant uterus. LM and FN are not made by the muscle layer of the uterus beneath the decidua which could explain why trophoblast does not normally invade beyond this point and rupture the uterus. This is an important mechanism used by the uterus to limit how deep trophoblast is allowed to go. Integrins are highly versatile molecules due to the way they are made. Depending on their structure, some bind preferentially to LM while others bind to FN. Binding to LM tends to make the cell stay still while binding to FN makes the cell move. In this way trophoblast, by expressing the right kind of integrins, and uterine decidua, by producing the appropriate LM or FN, can both exert control over implantation.

To make it easier for trophoblast to move through decidua, the ECM needs to be converted from its normally gluey consistency into a more liquid form so that the cells can push through it easily.[106] Trophoblast cells use a variety of enzymes to break down the protein structure of ECM, the most important of which belong to a family of 'matrix degrading metalloproteinases' (MMP). Trophoblast cells secrete these enzymes to digest their way through the ECM. This production can be easily demonstrated in the laboratory. When trophoblast cells are seeded onto culture dishes containing ECM, circular haloes of clear material representing digested matrix can be seen around each cell. Secretion of MMP is most pronounced by trophoblast cells from early pregnancy. This coincides with the period of greatest invasion during implantation. While trophoblast cells are secreting MMP, decidual cells counter this by producing inhibitors to these enzymes, appropriately known as 'tissue inhibitors of MMP' (TIMP). Interestingly, the gene controlling the production of TIMP is located on the female X chromosome. This is a further example of how the mother can exert a restraining influence on trophoblast invasiveness.

Trophoblast migration has often been compared to cancer cell invasion. Both use interaction of CAM with ECM to provide traction for movement. Both secrete MMP to digest their way through tissues. Both pregnant women and cancer patients produce TIMP to reduce the progression of trophoblast and cancer cells respectively. We have already referred to the ability of syncytiotrophoblast to penetrate blood vessels as cancer cells habitually do. We will encounter many more similarities in later chapters. There is a tendency to look at how trophoblast cells mimic the behaviour of cancer cells. It may be more profitable to reverse this comparison and ask why cancer cells behave like trophoblast. Cancer behaviour has not arisen *de novo* but is merely a re-expression of the attributes of a normal trophoblast cell but at an inappropriate time and place.

Interaction between integrins and ECM is not just something taking place mechanically outside the cell, it will also send signals into the trophoblast cell telling it how to behave. This mechanism of 'signal transduction' by which activities outside a cell transmit information to its interior shows how a cell can be influenced by its immediate surroundings. Two events happen inside the trophoblast cell. At first there is reorganization of its internal architecture leading to changes in cell shape which, in turn, influences its motility. Specific biochemical changes are then triggered inside the cell which determine its behaviour. Depending on the combination of integrins and ECM, the results can lead either to trophoblast cells becoming highly invasive or less invasive.

During implantation in normal pregnancy there is a switch from LM-binding integrins (non-motile) to FN-binding integrins (motile) on the surface of trophoblast cells as they migrate into decidua. In the placentas of women who have PE this switch does not occur.[107] This could explain the premature cessation of trophoblast invasion and inadequate arterial remodelling which characterize PE. The mechanism of healing, for example of a skin wound, closely parallels that of trophoblast migration during implantation. At the periphery of a wound the healthy skin cells are anchored by LM-binding integrins onto the underlying LM which is normally present in the deeper layer of the skin. During healing, the integrins on the new skin cells migrating to cover over the gap in the wound are transformed into FN-binding integrins. The wound itself contains FN derived from clotted blood instead of LM. The skin cells are now transformed from sessile to motile forms. These *in vivo* processes can be demonstrated by *in vitro* laboratory experiments. Trophoblast cells seeded onto a layer of LM or FN look different. Using time-lapse photography, trophoblast cells can also be shown to be more motile when cultured on FN than on LM. Such experiments confirm that the binding of integrins to LM tends to restrain trophoblast cells in a fixed position, whereas interaction with FN makes trophoblast cells more motile. This principle seems to apply also to other cell types such as skin cells. Here again, the similarity to cancer is striking. In laboratory cultures, cancer cells grown over a layer of FN survive and proliferate much better than those seeded onto uncoated culture dishes.

From the above description we can appreciate that the process of trophoblast migration into decidua is exquisitely controlled by a complex interplay of many components that include CAM, ECM, MMP, and TIMP. To complete the picture there is one more component we need to include. These are 'cytokines'. Cytokines[108] are soluble proteins secreted by cells that act on other cells nearby, providing a means of communication between them without having to make physical contact. Their production is often transient and the radius of activity is small so they tend only to affect cells in the immediate neighbourhood. This is known as 'paracrine' action. This is different from the 'endocrine' action of hormones which can influence cells distant from the source of production (see Chapter 7). Cytokines can sometimes also have an effect on the cell that produces them, in that the cell acts on itself. This is 'autocrine' action. Cytokines are very powerful and can exert their effects at extremely low concentrations. Only tiny amounts are needed which makes their presence in cells or tissues difficult to detect.

Both trophoblast and uterine decidua produce cytokines.[109] In this way they can influence each other's behaviour. Only those cells with the appropriate receptors can respond to a particular cytokine. This serves to restrict the cytokine's activity and confers what biologists refer to as specificity. Trophoblast cells have surface receptors for a variety of cytokines secreted by uterine cells so they are potentially capable of responding to these signalling proteins from the mother. Furthermore, different trophoblast populations at the implantation site have their own array of cytokine receptors, so that they are responsive to distinct groups of cytokines. Rapidly proliferating trophoblast cells have receptors for growth-promoting cytokines while those involved with invasion have receptors for cytokines that will affect their migration. For example, the cytokine known as transforming growth factor beta (TGF-beta) has been shown to alter the levels of integrins on the surface of trophoblast cells and can also affect the production of TIMP. Another cytokine called leukaemia inhibitory factor (LIF) is reported to decrease the production of MMP by trophoblast. These cytokines will clearly influence implantation. Both TGF-beta and LIF are produced by cells in the decidua, so here is another way for the mother to control trophoblast invasion.

While these observations provide a good case for cytokines playing an important role in implantation, that they actually do so *in vivo* has been difficult to prove. One reason is that the function of one cytokine can be mimicked by another, a phenomenon known as 'redundancy', so that when one is removed another takes over.

This is a good back-up strategy; if plan A does not work, use plan B. This makes defining a role for a particular cytokine somewhat problematic. The most convincing evidence demonstrating the importance of a cytokine in implantation *in vivo* is provided by an elegant knock out (KO) experiment in mice where the gene for LIF production is artificially rendered non-functional. Female KO mice lacking the LIF gene are unable to support implantation of their embryos, but if these embryos are transferred to a normal mouse they will implant. Furthermore, infusion of LIF back into the KO female will restore her ability to implant. In KO female mice in which the gene for the LIF receptor (LIF-R) is deleted instead of the gene for LIF itself, their embryos can implant but they die within 24 hours of birth from poorly developed placentas. LIF seems to have a dual role to play in reproduction—one in placental growth and another in implantation.

The curious reader might wonder why these cytokines have been given names which, at first sight, have nothing to do with trophoblast or decidual function. They were originally identified by their actions on blood-forming cells in the bone marrow and are described according to their effects on these cells. It was only later that these cytokines were shown to influence other cell types besides blood cells, but their original names remained. Many cytokines have these kinds of pleiotropic or multiple functions. This characteristic, together with redundancy mentioned earlier, further complicates the evaluation of cytokine bioactivity in implantation.

There is one more interaction which I will now introduce briefly. This is between trophoblast and a population of cells called Natural Killer (NK) cells present in decidua. These cells are from two different individuals: placental trophoblast cells are derived from the baby while decidual NK cells belong to the mother. What happens when these incompatible cell types meet? We shall focus on this question in Chapter 6. It is one of the great mysteries of pregnancy.

What we have described so far for placental implantation has relevance also for general cell biology. A cell is not a totally independent entity. Although the properties of each cell are pre-determined by its genetic recipe, its behaviour is modified by interaction with neighbouring cells and with the immediate surroundings. Trophoblast cells are good examples. They are inherently programmed to proliferate and spread. They will behave sensibly while inside the uterus because of the restraining influence of decidua. But in ectopic situations, where there is no decidua, their true aggressive nature is revealed. Like William Golding's *Lord of the Flies*, children without parental discipline will revert to type and become savages. The same goes for cells during early embryonic development. They will grow uncontrollably unless something reins them in. Cancer biologists Carlos Sonnenschein and Ana Sato in their book *Society of Cells* view cancer as embryonic development gone awry.

After implantation, the rapidly expanding trophoblast cells eventually form buds or protrusions over the whole surface of the placenta at the implantation site. These are called 'villi' (also referred to as the 'villous tree' because of the way they branch), many of which dip into the pools of maternal blood which have been created by the initial syncytiotrophoblast invasion as described earlier. This appearance has been likened to a 'mop in a bucket of blood'. Although somewhat gruesome, this description does convey an accurate image of what is happening. At first the villi are composed of solid clumps of trophoblast only but by the end of the third week new blood vessels begin to appear in the core of these villi. These new blood vessels arise from primitive cells within the core of the villi and are derived from the fetus. They first differentiate into endothelial cells which will form the lining of the future blood vessels. The endothelial cells then organize themselves into tubes and, by the ninth week of gestation, will fuse with each other to become complete vessels. This is the beginning of a functioning vascular network. This whole process is known as 'angiogenesis', which describes

how new blood vessels are formed.[110] These events can be reproduced in the laboratory. Endothelial cells isolated from the umbilical
cord placed in culture can be seen to form tube-like structures. This
will happen only when the endothelial cells are incubated in a low
oxygen environment. We have seen earlier in this chapter that this is
the kind of hypoxic environment created by trophoblast plugs
blocking the mouths of the uterine arteries during the first trimester
of pregnancy. Their influence on placental angiogenesis shows what
an important task these plugs perform. Hypoxia also triggers angiogenesis in tumours, a further example of placenta and cancer development following similar pathways.

A variety of growth factors are involved in this process, some are
pro-angiogenic while others are anti-angiogenic. Together, they
provide proper balance and control. Some factors influence the early
stages, such as the initial formation of endothelial tubes, while
others come into play later, like prolonging the survival of endothelial cells or maintaining the integrity of the blood vessel walls once
they are made. The best understood of these growth factors is the
vascular endothelial growth factor (VEGF) and its corresponding receptor VEGF-R. The receptor acts to absorb out free VEGF which
thus functions as an anti-angiogenic factor by inhibiting the angiogenic activity of VEGF itself. The importance of VEGF can be demonstrated in mice by artificially disrupting the gene controlling its
production. Embryos in these VEGF (KO) mice do not survive and
the blood vessels in their placentas are highly disorganized.

The newly made blood vessels in the placental villi soon connect
up with larger vessels of the umbilicus, which in turn become united
with the blood vessels coming up from the embryo itself. A circulation is now formed in which blood passes from the embryo to the
placental villi, travelling by the umbilical artery and returning to the
embryo via the umbilical vein. This blood is circulated in gradually
increasing volume by the embryo's heart, which begins to beat around
the fourth week of gestation. Through the layer of trophoblast lining

the villi, exchange is made between the pools of maternal blood outside the villi in the intervillous space and the embryonic circulation inside the villi. In this way the placenta acts as the lungs, kidneys, and digestive system for the embryo (from 3–8 weeks) and its subsequent development into a fetus (after 8 weeks).

Blood flow through the placenta is unusual. Arteries supplying oxygenated blood to organs normally divide into a network of small capillaries as they reach the tissues. It is here that exchange takes place between blood and tissues. The used blood is then collected up by veins and returned to the circulation. This is not what happens in the placenta. There are no capillaries. Instead, the incoming uterine arteries empty directly into the pool of maternal blood accumulated in the intervillous space, described earlier as the 'bucket of blood' into which dip the placental villi. The result is a dramatic fall in blood pressure from 70–80mm Hg at the end of the uterine artery to around 10mm Hg in the intervillous space, like a river flowing into a large lake. All the force is spent. Blood then passes to the uterine veins from where it drains back into the maternal circulation. The blood flow within the intervillous space is, thus, rather sluggish. But the blood does move, albeit at a slow pace. This is good for efficient exchange. Washed over gently by this slow-moving blood, the trophoblast cells covering the villi have ample time to absorb nutrients and oxygen while getting rid of waste products from the embryo at the same time. In his book on *Comparative Placentation*, Peter Wooding maintains that flow rate is probably more important than concentration gradient, or even the number of barriers within the placenta, in determining the efficiency of transfer between fetal and maternal circulations.

Although fetal blood is now placed in close juxtaposition with maternal blood, it must be emphasized that there is no *direct* communication between the two circulations. They are entirely independent systems so the two bloods never mix. It was the Greek physician Galen (AD 130–200) in his early studies of the placenta who started the confusion by claiming that there was a connection. Galen was an authoritative figure throughout the Middle Ages, and because he was held in such high esteem his incorrect conclusion remained unchallenged for over a thousand years. It was not until the Renaissance period that an Italian physician, Giulio Cesare Arantius, professor of medicine and anatomy at the University of Bologna, (1564) dared to correct Galen's mistake in his treatise *De Humano Fetu* and announced to an astonished scientific community that the embryonic and maternal circulations in the placenta were entirely separate and there were no connections between the two. This conclusion was subsequently confirmed by investigators from a more modern era, such as William Harvey and the Hunter brothers in the 17th and 18th centuries by the injection of coloured dyes into the placental circulation followed by careful dissection to see where they led. The advent of the microscope in the 19th century allowed further progress in understanding the placental circulation. Of course, mixing of fetal and maternal blood must not be allowed. Otherwise, there would be transfusion of blood between the baby and mother, with all the attendant hazards. This is obvious to us now but it wasn't in Galen's day. The vascular system of the placenta is really a very ingenious architectural design, allowing two blood circulations to come into intimate contact to facilitate exchange and yet without permitting them to mix. The one situation when direct mixing of blood from two individuals occurs within the placenta is with monochorionic twins. Their circulations occasionally join up leading to what is known as 'transfusion syndrome'. Either one or both twins may succumb.

The intervillous space itself is rather curious. Blood normally never leaves the confines of the vascular system. It remains enclosed within the system throughout its circulation around the body, from the heart to blood vessels returning to the heart via the lungs. Blood leaks out only when there is injury, as we have all experienced with cuts and bruises. But here within the intervillous space there is a pool of maternal blood lying *outside* her vascular system. The space is not lined by the usual layer of endothelium but by trophoblast, a tissue supplied by the fetus and not by the mother. We noted in an earlier chapter that syncytiotrophoblast is well-suited for this task because its surface has a non-stick, Teflon-like property that prevents blood from clotting. For a considerable time, anatomists found it difficult to come to terms with this peculiar arrangement. The intervillous space is unique.

At birth, the baby enters a new world. Blood is diverted from the placenta to the lungs. Resistance to blood flow in the fetal lungs is high at first because the lungs are collapsed until the first breath is taken and the lungs expand. Thereafter, resistance falls and blood can then pass freely through the lungs, allowing proper oxygenation of the blood. This transformation from placental to pulmonary respiration in the baby takes place over several minutes. The age-old controversy that has been raging since the days of Aristotle as to the best time to clamp the umbilical cord after delivery of the baby stems from this. From physiological considerations, it would seem that a few minutes of delay rather than immediate clamping of the cord should be favoured to allow time for this transformation to take place. Aristotle himself approved of the practice of not cutting the cord until the placenta was delivered, after he noticed that midwives frequently squeezed blood remaining in the cord back to a newborn who needed to be revived. A significant volume of blood can be retained in the cord which should be restored to the newborn. Cord blood has become an important source for harvesting stem cells that can be used to repair damaged adult tissues (see Chapter 9). Those involved would obviously like as much blood as possible to be left in the cord. The entry of commercial interest has changed the terms of the debate.

6

Nature's Transplant

People often ask me about the area of my research. When I reply that I am a reproductive immunologist, the reaction is usually one of puzzlement. This is understandable. The term immunity (Latin *immunis* = exempt) was coined from the observation that individuals who had recovered from certain infections such as chickenpox or measles were thereafter protected from the disease. We are all familiar with this phenomenon. From this originated the science of immunology which was, therefore, mainly concerned with how the body defended itself against infections. So what has immunology got to do with reproduction? Quite a lot, as this chapter sets out to explain.[111]

Our immune system protects us from diseases by destroying any pathogenic organisms which may have gained entry into our body.[112] It recognizes these organisms as not being part of ourselves, and so to be eliminated. This discrimination does not apply only to invading pathogens. Cells, tissues, and all bodily parts from another individual are also foreign entities because they are not our own. Except for identical twins, we are all genetically different from each other. The immune system can detect these differences, a process known as 'allorecognition' (the prefix 'allo' meaning from another individual), and that is why organ transplants from one person to another are usually rejected. The immune system has an uncanny ability to discern host from unwelcome visitors. As Macfarlane Burnet elegantly phrased it in the early 1940s, the immune system can discriminate 'self' from that which is 'non-self'. The *Oxford English Dictionary* defines 'self' as 'a person's essential being that distinguishes them from other people'. This seems an appropriate description of how the immune system works. The elusive nature of 'self' has long been a favourite topic for philosophical discourse. Psychology, neurology, sociology, and even religion have all contributed their own versions of selfhood, consciousness, and individuality. The introduction of the immunological 'self' has at least provided a more solid biological basis for this hitherto rather nebulous concept.[113] Prior to Burnet, the term 'self' was not used in immunological literature, but this metaphor is now well and truly established in the lexicon of immunology.

How does the immune system make this distinction? First of all it has to learn to recognize the unfamiliar features of the 'non-self'. Once learned, it can remember these features so it will make a better and faster response with the next and subsequent encounters. This response is directed specifically only towards the 'non-self' it has learnt to recognize and to no other. These are the three main characteristics of what is known as the adaptive immune system which we use for defence: learning, memory, and specificity. The adaptive immune system is so named because it is designed exclusively to fit just one particular 'non-self' target, a kind of bespoke tailoring and, like any good quality made-to-measure product, it requires a bit of time to construct. Our response to infections, for example, takes at least a few days.

In addition to the adaptive immunity just described, we possess another immune system known as innate immunity. The innate system is simpler and less sophisticated than the adaptive one. It is a more ancient system in evolutionary terms, already in use by invertebrates a long time ago. Adaptive immunity, in contrast, is a more recent vertebrate invention, appearing only since the emergence of cartilaginous fishes such as the shark. But the 'old' system has not been entirely swept away by the 'new'. Some aspects of the old system have been retained and now continue to function as the innate immunity of vertebrates, side by side with adaptive immunity.

In innate immunity, no prior learning is required, unlike adaptive immunity when new knowledge about the unfamiliar pathogen or transplant has to be acquired. This observation prompted Burnet to conclude that discrimination in innate immunity is based on 'self' recognition rather than on 'non-self'. Any foreign cell that lacks 'self' markers will be identified immediately, so this system primarily detects the *absence* of 'self', or more succinctly, 'missing-self'. There is no need for a period of learning because the innate immune system already knows what 'self' looks like. Using 'self' as a template, any foreign cell that does not conform to this image will be rejected. There are then two ways by which our immune systems recognize something as being foreign. One is when the foreign cell possesses molecules which the host does not have (presence of non-self). This is the detection system used by the adaptive immune response. The other is when a foreign cell does not have the molecules which the host possesses (absence of self). This is used by the innate immune response.

Our ancestral innate immune system might have been designed primarily to survey and preserve the organism's internal integrity, what physiologists call 'homeostasis'. Its house-keeping role here is to eliminate effete and senile cells within the body as well as guard against rogue cells that might transform into cancer. In complex multicellular creatures like ourselves there will always be disharmony among cells within the body and it is the responsibility of the innate immune system not to let this get out of hand. Defence against external aggressors by the adaptive immune response evolved later. This view of a dual role for the immune system was expounded by the Russian biologist Elie Metchnikoff over 100 years ago. Between our innate and adaptive immune systems, we can deal with just about any kind of insult arising either from within or from without. Which of these two systems is used by the mother to recognize her placenta? We shall see.

In the early 1950s, Peter Medawar established that rejection of an organ, such as a kidney, transplanted from one person to another was due to genetic differences between donor and recipient, the donated organ being recognized as 'non-self' by the recipient's adaptive immune response.[114] He was awarded the Nobel Prize for this discovery. At the same time, he posed the question of why then the mammalian fetus is not similarly rejected, since it contains genes from the father and should, therefore, be regarded as 'non-self' by the mother. If a woman were to receive, for example, a kidney donated by her husband, she would reject it. Yet the fetus, containing similar genes from her husband, is readily accepted. We have the intriguing situation in which a foreign tissue (fetus) not only survives but continues to grow within a host (mother) without rejection. This is the 'immunological paradox' of pregnancy. The analogy of the 'fetus as an allograft' has become a popular slogan in reproductive immunology but a more critical examination reveals that this analogy may not be as appropriate as it initially appeared to be. As we have seen in an earlier chapter, the fetus itself, although sitting inside the uterus throughout pregnancy, does not make direct contact with the mother at any time. All communication between the fetus and mother is made via the placenta. In the context of immune recognition, it is how the mother sees her placenta that counts. The placenta, though, is not an exact replica of an organ transplant. To the mother, the placenta is certainly 'non-self' because the father's genes are present. But it is only half 'non-self', in that the mother's own genes contribute to the other half. These maternal genes will give the mother a comforting sense of 'self'. So we have two opposing forces at work here: the father's foreign genes could provoke rejection by the mother's adaptive immune response while the presence of the mother's own genes would stop her innate immune response being triggered by 'missing-self'. The potential for destruction and protection of the placenta is present at the same time. Later, we shall examine how the placenta negotiates this complex immunological maze.

There are occasions when the mother's own genes are not present in the pregnancy. This occurs when the egg is from another person (donated egg) or when the fertilized egg is transferred to a third-party recipient (surrogate mother). In both these cases, the developing egg in the womb is unrelated to the mother carrying the pregnancy. Without the protection provided by the mother's genes, problems do arise. There are two types of surrogacy. 'Traditional' surrogacy is when the egg is from the surrogate mother and the fertilizing sperm from the commissioning father. 'Gestational' surrogacy is when both the egg and sperm are from the commissioning couple. The success rate is believed to be lower in 'gestational' surrogacy. This could be due to 'absent self' in the conceptus because maternal genes from the surrogate mother will be totally absent. In egg donation, the frequency of the mother developing pre-eclampsia can be as high as 30 per cent compared to approximately 1 in 2000 in normal pregnancies. Again, this increase can be explained by the total absence of the mother's genes in the placenta.

What are the genes that make us so different from one another? They are a group of genes known as the *Major Histocompatibility Complex*, usually just referred to by its acronym, MHC. It is the largest family of genes in the human genome. All animals have their own MHC. In humans, a group of genes within the complex are called *Human Leukocyte Antigens* or *HLA*. The products of these *HLA* genes are found on the surface of all our cells and tissues, conferring on them a unique HLA profile, which forms the blueprint for our individuality and defines 'self'. This is the major hurdle in clinical transplantation. Hence, the initial requirement before transplantation is to try to 'match' up as closely as possible the HLA of the donor with that of the recipient in the hope that this will reduce the severity of the rejection. The closer the match, the less 'non-self' the transplant is perceived to be by the recipient. HLA are the most highly variable molecules in the body. Many genes go towards making them (polygenic). Once made, the resultant molecules are different from each other (polymorphic). Furthermore, each individual has inherited genes from both parents (heterozygous). It is not too difficult to visualize how this combination of polygeny, polymorphism, and heterozygosity will allow a vast number of different permutations and hence an enormous variety of HLA molecules to be generated. As a comparison, most of us are familiar with the ABO blood groups. An individual can be either group A, B, AB, or O, so there are only four possible variations. In contrast, it has been calculated that there are in excess of 10^{12} or 1 million million different HLA. Because of this, within the human population, no individual will have the same HLA.

Why do we need such a huge degree of variability? We know that HLA molecules are responsible for graft rejection because they distinguish 'self' from 'non-self'. But this, actually, is not their function. It is a consequence of their function, which is to defend us against pathogenic or disease-causing organisms. HLA molecules are a crucial component of our immune defence system. They work by informing our immune system that we are being invaded by pathogens and need to trigger an appropriate defensive response. In order for a pathogen to be recognized by the host to start the immune response, the pathogen has to be broken down into small pieces, or peptides, which are then presented on the cell surface by binding onto tiny grooves in the HLA molecules. Some HLA grooves fit better with peptides from a particular pathogenic organism, making them more efficient in starting the immune response to these pathogens. In this way, the pattern of HLA molecules which an individual possesses will influence his or her resistance to a particular disease. For example, the HLA designated as B53 protects against the disease malaria, so that over many generations this HLA has become common in West Africa, where malaria has long been endemic. Otherwise, HLA-B53 is rare in other parts of the world. Because there is an almost infinite variety of pathogens which we are likely to encounter, we in turn need an extensive repertoire of HLA molecules with different grooves ready to deal with them. If an epidemic of a particular disease were to occur, there will always be some individuals within the population with the appropriate HLA who will be resistant to that infection. This will guard against the whole population being wiped out by an epidemic. The evolutionary pressure to maintain a wide spectrum of HLA molecules within a population is to protect us against diseases. It has not evolved to frustrate transplant surgeons, although this is the undesirable consequence.

How then does the placenta with 'foreign' paternal HLA avoid rejection by the mother's immune system? Reproductive biologists have tried to find a unifying concept to explain the immunological paradox of placental survival. Scientists are invariably fascinated by a paradox. It invites questions and questions require answers. Is the mother immunologically ignorant of her placenta or does she tolerate its presence? The two are not the same. In the former the mother is unaware of the presence of her alien placenta, while in the latter she is aware but does not respond. There is also a third possibility in that the mother is aware and does respond but in a totally unexpected manner. This last mechanism is the most intriguing. We shall look at this in more detail later.

In the placenta, the layer of syncytiotrophoblast which lies in direct contact with the mother's blood and, therefore, is at the mercy of the full force of her immune system, does not have any HLA at all on its surface. Syncytiotrophoblast is the only cell in the body which is devoid of HLA. In this way, it has cleverly cloaked its 'non-self' nature by becoming immunologically invisible to the mother. This is how it evades immune rejection by the mother. From this observation, the immunological paradox of pregnancy was considered solved. Unfortunately, a spoke in the wheel was introduced when it was discovered that the other trophoblast population that invades into the uterus during implantation, such as interstitial trophoblast (see Chapter 5), does have HLA molecules. Curiously, these trophoblast HLA are not the same as those normally found on other cells of the body.[115] The family of HLA molecules are designated by different letters of the alphabet. All cell types in the body show or express three of these, namely HLA-A, -B, and -C. In contrast, trophoblast does not have HLA-A or -B. Instead, it expresses HLA-C, -E, and -G. HLA-A, -B, and -C are the molecules principally responsible for graft rejection because they are highly polymorphic and are most likely to differ between donor and recipient and, hence, to trigger rejection. It is easy to understand why trophoblast cells should want to discard those -A and -B molecules that would identify them as 'non-self' by the mother, but why acquire another set with the unusual combination of -C, -E, and -G? If evasion of maternal rejection is all that is required, then surely the simplest solution is for all trophoblast populations to be like syncytiotrophoblast and have no HLA at all?

Unlike the other HLA molecules, -E and -G are non-polymorphic, that is to say everyone's -E and -G are the same. Such molecules have always been a puzzle to immunologists.[116] As we discussed earlier, the purpose of having a wide variety of different HLA in the population is to ensure there is always someone with the appropriate HLA molecule that combines efficiently with the peptide from a particular pathogen and is resistant to that organism. Since -E and -G are the same in everyone, they are obviously not for defence against disease. So what are they for? There was a time when HLA-G was dismissed disparagingly as a vestigial molecule left behind by evolution with no purpose in life any longer. It represents a gene in decline, a gene which has passed its sell-by-date, with the evocative description by Peter Parham of it being 'a rotting hulk in an evolutionary junkyard'. But HLA-G is making a comeback. There is a view now that it may have assumed other functions other than defence against infections, some of which may be related to reproduction. Indeed, HLA-G is unique in being present only on placental trophoblast cells and not found on any other cells of the body.[117] It is strongly expressed by trophoblast in the first trimester of pregnancy and then gradually wanes during the later stages. This restriction to a single tissue, and changes in expression at different stages of pregnancy, point to HLA-G playing some role associated with trophoblast. Of the non-human primates studied, Rhesus monkeys and baboons also have a molecule on their trophoblast that exhibits many of the characteristics of human HLA-G.[118] This conservation across species lends strong support to the idea that HLA-G is likely to be functional. Molecules that are useless tend gradually to disappear during evolution and not be conserved. What might this function be?

We will speculate on the function of HLA-G later. At this point, it would be useful to revisit Chapter 5 on 'Implantation'. Trophoblast cells must be allowed to invade into the uterine decidua in order to transform the blood vessels there. Otherwise, the blood flow to the fetus and placenta will be insufficient to sustain their development, resulting in a spectrum of disorders, such as recurrent miscarriage, fetal growth retardation, and pre-eclampsia. At the same time trophoblast must not be allowed to penetrate too far into the uterus with the attendant danger of placenta creta and uterine rupture. The situation in implantation then, is more subtle than that of a conventional organ transplant. Implantation requires a *balanced* environment that allows trophoblast invasion but, at the same time, curbs excessive penetration, which could endanger the mother. It is not just a question of acceptance or rejection by the mother as in an organ transplant. Instead, a state of peaceful coexistence between the placenta and the mother needs to be achieved. This requires a special kind of immune mechanism. How this system might work is slowly being unravelled.

The immune cells that participate in graft rejection belong to a family of lymphocytes called T cells and B cells. They infiltrate the area surrounding a transplanted organ. The cells in the uterus around the implanting placenta are not T and B cells but are made up mainly of another family of lymphoid cells mentioned earlier, the Natural Killer (NK) cells.[119] This further reinforces the view that the maternal immune response to the placenta is not the same as that seen in a recipient of a conventional organ graft. During the evolution of our immune defence system, NK cells emerged long before T and B cells. NK cells were already present in our invertebrate ancestors and have persisted to the present day to form the innate immune system of vertebrates. T and B cells of the more modern adaptive immune system responsible for graft rejection evolved later. They are relatively recent arrivals.

It is worth stressing that pregnancy represents the only *naturally* occurring situation when tissues from two genetically distinct individuals come together. Clinical organ transplantation is an *artificial* situation created by advances in medical techniques. It is not a natural phenomenon. Way back in evolutionary time, however, in their crowded marine environment, direct cellular contact between genetically distinct sedentary invertebrate colonies frequently occurred. To guard against fusion or contamination by members of alien colonies, the ability to recognize 'non-self', or allorecognition, developed naturally to combat this threat. In higher vertebrates, this form of natural allorecognition contact has largely disappeared and is now encountered only in pregnancy. It should come as no surprise then that the immunological relationship between the placenta and mother during pregnancy utilizes a more primitive innate immune defence system whose characteristics are more akin to those between unrelated invertebrates in the sea than between graft and recipient, as seen in contemporary organ transplantation.

As already described in Chapter 5, trophoblast cells normally invade the uterus up to a certain depth and no further. A definite boundary is formed between the invading placenta and the uterus. This represents an area of mutual demarcation of territories between two genetically dissimilar tissues: placenta and mother. There is no destruction of either of the two. This is also what happens in invertebrates. Experiments with sponges have shown that when two genetically unrelated sponge colonies are put together, this is not necessarily followed by rejection of one by the other. Instead, a barrier is formed between the two colonies. This bears a remarkable resemblance to what happens at the meeting point between the placenta and the uterus. It is territorial demarcation rather than rejection.

How might this innate immune system work in pregnancy? A great deal of attention is directed towards uterine NK cells.[120] Similar NK cells are found in the uterus of all species so far studied, including those with non-invasive epitheliochorial placentas, such as pigs, horses, and cattle, so these cells are highly conserved across species despite differences in placental type.[121] In the human uterus, NK cells are sparse in the early phase of the menstrual cycle. They begin to increase in numbers in the later stage of the cycle and remain abundant in the pregnant decidua.[122] At the implantation site, NK cells congregate around the invading trophoblast cells of the placenta. The influx of NK cells into the uterus, therefore, correlates in both time and place with the implanting placenta. It is this temporal and spatial relationship between uterine NK cells and invading trophoblast that has led to the idea that the former could exert some controlling influence on the latter. There is one thing that is puzzling. We see from Figure 13 that these NK cells already begin to

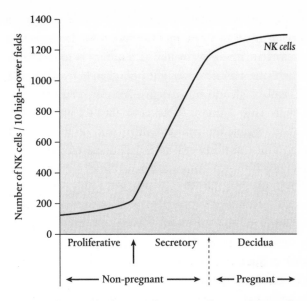

FIGURE 13 Graph showing the accumulation of NK cells throughout the menstrual cycle and in pregnancy.

appear in the uterus during the later stages of the menstrual cycle even before pregnancy occurs. The placenta, then, cannot be held responsible for recruiting these cells to the uterus. So, if we are to propose that uterine NK cells are there to control trophoblast invasion, then we have to assume that they are already committed to this role even before they have a chance to encounter the alien or 'non-self' placenta. This is very different from what happens in organ transplantation. Immune cells, such as T and B cells, start to accumulate only when the transplanted organ is in place but not before.

Perhaps uterine NK cells have an additional function to perform before they meet up with trophoblast. If pregnancy does not occur, NK cells begin to die in the last few days of the menstrual cycle. In fact, NK cell death is the first detectable sign that menstruation is about to start. NK cells are known to produce a variety of growth factors, some of which might be important in maintaining the uterine endometrial lining in a healthy state. With the demise of NK cells these factors are no longer available, resulting in the breakdown of the endometrium which is sloughed off as menstrual bleeding. The life and death of NK cells are, therefore, intimately related to the menstrual cycle. Their primary function might be a physiological one, which is to regulate the changes taking place in the non-pregnant endometrium. This fits in well with the homeostatic view proposed by Metchnikoff that an important function of the innate immune system is to preserve the internal integrity of the body. If pregnancy takes place these NK cells then take on an additional immunological role to control trophoblast. One cell doing two different jobs depending on circumstances is an excellent demonstration of flexi-working and a fine example of functional economy in biology.

No other organs of the body are populated by NK cells as in the uterus. NK cells are present in blood but these have different characteristics to uterine NK cells, indicating that the two populations are either of diverse origins or are at different stages of development. This fact is often overlooked by clinicians dealing with infertility problems, leading to claims that monitoring variations in blood NK cells mirrors what is happening within the uterus. This conclusion is unwarranted.

How NK cells work has long been a mystery. NK cells are seen to preferentially kill target cells which have low or absent HLA molecules on their surface. Their recognition system, therefore, is based on detection of the absence of 'self' or 'missing-self', as already described for innate immunity.[123] NK cells, then, operate in exactly the opposite manner to T cells in graft rejection, which kill 'non-self' cells of the foreign graft. While T cells of adaptive immunity recognize the presence of differences between graft and host, NK cells in innate immunity are alerted to the absence of similarities. Our defence system is really rather efficient in that it has two complementary mechanisms for identifying whether something is foreign. One of the characteristics of cancer cells and cells infected by viruses is that they have low or absent HLA on their cell surface. This makes them 'foreign' in the eyes of NK cells because they are 'missing-self' due to the lack of HLA. So, besides their function in reproduction, NK cells are also the front line of defence against virus infections and will eliminate developing cancer cells within the body. This latter function again fulfils the internal surveillance role of the innate immune system envisaged by Metchnikoff.

The 'missing-self' hypothesis for NK cell function is a good illustration of how scientific ideas develop. During the period of the Cold War, the Swedish coastline was frequently invaded by numerous spying submarines from different nations. The Swedish government encouraged the population to report sightings of these foreign submarines by issuing a booklet with illustrations of what these foreign submarines looked like. With commendable lateral thinking, some bright official suggested that instead of producing a series of pictures showing foreign submarines, why not issue just one picture of a Swedish submarine with the request that people should only report if they were to see a submarine that did *not* look like Sweden's own submarine. A Swedish scientist, Klaus Kärre, noticed this and formulated the 'missing-self' hypothesis for NK cell immune recognition.

While the recognition of 'missing-self' is a good basic model to illustrate NK cell function, it is not the complete story. They are much cleverer than we thought. We now know that NK cells do not merely detect the absence of HLA, they also possess a variety of receptors for HLA, including HLA-C, -E, and -G which are present on trophoblast so they are potentially capable of discriminating between different trophoblast HLA molecules, just like T cells can except NK cells do it in a different way. NK cells, therefore, have a dual capability: they can detect 'non-self' as well as 'missing-self'. To add to the complexity, there are two kinds of NK cell receptors, some of which are 'activating' while others are 'inhibitory'. That is to say, some receptors 'switch on' while others 'switch off' NK cell activity after encountering HLA, so the final outcome is dependent on which receptors are engaged. The original description of these cells as 'Natural Killer' cells is misleading because the term conjures up an image of NK cells relentlessly setting out to attack and kill their targets. This is not always the case. NK cells are more versatile than mere killing machines. They produce a variety of substances with multiple functions, some of which might even have beneficial effects on trophoblast. These functions could be more relevant to implantation than killing. We noted earlier that successful implantation requires a *balanced* response by the mother towards the placenta. The presence of receptors that can either activate or inhibit uterine NK cell activity could provide just the right degree of balance.

The family of NK cell receptors for HLA-E are all inhibitory, so when uterine NK cells encounter this molecule on trophoblast they will be 'switched off'. In this way, HLA-E would protect trophoblast by preventing uterine NK cells from killing it, and this is the reason for its presence on trophoblast. The part played by HLA-G is less clear. It is a molecule that still has not been assigned a function. The reason for this is that the identity of the NK cell receptor for HLA-G has proved somewhat elusive.[124] An HLA molecule fixes onto its own specific receptor, in the manner of a lock and key, in order to trigger a reaction. Without precise knowledge of its receptor, it is difficult to fathom how HLA-G might work. Since HLA-G is present only on trophoblast, the signal it sends to the mother must be a 'trophoblast' signal of some kind and not a paternal signal, because HLA-G is non-polymorphic and will be the same in both father and mother. What might this signal be? At present we are not sure. It could be argued (albeit teleologically) that since HLA-G is a trophoblast signal, the response it wants to provoke in the mother is one that will be beneficial to the survival of the placenta. There is evidence that points to the receptor for HLA-G being an activating receptor. But triggering this receptor does not provoke the NK cell to reject the placenta. Instead, it stimulates the NK cell to produce some substance that might enhance the growth of the placenta.[125] Uterine NK cells are known to secrete growth factors that can influence the development of new blood vessels, a process known as angiogenesis, thereby providing a mechanism for the placenta to increase its own blood supply. If this is indeed what happens, it is really quite a smart move by the placenta to utilize the mother's NK cells for its own benefit: a good example of placental–maternal cooperation.

The formation of new blood vessels is a recurrent theme in cell biology because the process is constantly required for the regeneration and repair of tissues throughout the life of an individual. This is seen also in the growth of tumours. Cancer cells proliferate so rapidly that they quickly outstrip their blood supply, unless they can extend their vascular network the tumours will die. Many drugs used to treat cancer are designed to block the development of new blood vessels. Although HLA-G has not been detected in other normal cells besides trophoblast, there are reports that HLA-G may be expressed in some cancer cells. Does HLA-G have a similar angiogenic role to play in cancer as it has in the placenta? This has exciting implications for cancer biology. Other chapters have frequently drawn attention to shared characteristics between trophoblast and cancer cells. The expression of HLA-G may be a further example of this similarity. Because of its unusual nature and poorly understood function, HLA-G continues to attract an enthusiastic following, so much so that an international scientific meeting is convened in Paris every other year devoted entirely to this molecule.[126]

Although the function of HLA-G is not known it is, nevertheless, a very useful molecule. Since HLA-G is present only on trophoblast, it is a reliable marker by which to distinguish trophoblast from other cell types. To take advantage of this property, the first requirement is to find a way to detect the presence of HLA-G. The usual approach is to make a specific antibody to HLA-G by injecting the molecule into an animal, such as the mouse. Since HLA-G is not a normal constituent of the mouse, it will be regarded as 'foreign' and the animal's immune response will make an antibody against it, in the same way as we make antibodies against alien invaders. This antibody will seek out HLA-G and bind to it. To make this union visible, the antibody is usually coupled to a coloured dye. Trophoblast cells expressing HLA-G will be stained and can then be identified.

Of course, the reliability of this detection system is entirely dependent on the specificity of the antibody, that is to say whether the antibody is directed exclusively at HLA-G and to nothing else. Otherwise there will be cross-reactions and results become inconclusive. This is where problems arise and is the major source of controversies. Over 80 per cent of the HLA-G molecule is identical to other HLA, such as -A, -B, or -C, so that when injected into a mouse the animal will make antibodies directed at all the common components of the HLA molecule and not just to HLA-G. These antibodies will not have the required specificity for HLA-G because they will cross-react with all other HLA molecules. They are known as 'pan' HLA antibodies. They can be useful for determining whether a cell is HLA positive or HLA negative but they cannot discriminate between the different HLA types. So how do we go about making an antibody directed only at the remaining 20 per cent of the HLA molecule which is unique to HLA-G? My own laboratory spent a considerable amount of time devising a method to circumvent this problem.[127] We created 'transgenic' mice and used them as recipients for antibody production. These are mice whose genetic makeup has been artificially modified to contain genes for the common parts of the HLA molecule. These transgenic mice, when injected with HLA-G, will not react against the common components of the HLA molecule because they will be viewed as 'self'. Instead, they will produce antibodies only to that 20 per cent part of the HLA molecule which is unique to HLA-G, thus generating a HLA-G specific antibody that will not cross-react with other types of HLA.

The procedures are highly complex and very time consuming. But the efforts are worthwhile because the antibody is a powerful tool for answering a variety of questions about the biology of trophoblast. For example, at the implantation site where the placenta attaches itself to the uterus, trophoblast cells are so intimately mixed with other uterine components as to become invisible. Appearance alone is insufficiently distinctive to distinguish between various cell types because cells tend to look very much alike under a microscope. The availability of a specific trophoblast marker, such as HLA-G, has made the picture much clearer. We can now visualize precisely where trophoblast cells are located at the implantation site, how deeply they have invaded into the uterus, the manner by which they have modified the uterine spiral arteries, and the close contact they make with uterine NK cells. These are all features of the implantation site already described and drawn together in Chapter 5. The reader will now appreciate how this picture is made up and how much background work has gone into creating it.

Such a picture of the implantation site is certainly worth 'more than a thousand words'. Nevertheless, it remains just a static image representing one point in time, like a still photograph. 'Like songs of love, they much describe they nothing prove' (Matthew Prior 1687–1729). I first heard this quotation many years ago in a lecture given by the Cambridge physiologist Peter Wooding in Las Vegas (yes, at that mecca of gambling!), comparing the placental structure of different animals. I have always remembered the quotation because it so elegantly sums up the foremost inadequacy of histology which shows us what happens at the implantation site, but cannot tell us how it happens. To answer this second question, it is necessary to extract trophoblast cells from the implantation site and to design appropriate experiments in the laboratory. An antibody to HLA-G can help to do this.

Since an antibody to HLA-G can be used to identify trophoblast, it follows that it can also be a means of isolating these cells. To obtain a stock of pure trophoblast cells for experiments is a vital starting point for research on the human placenta. This is when a machine called the Fluorescence Activated Cell Sorter (FACS) comes into play. The HLA-G antibody is first conjugated to a dye which fluoresces in the presence of ultraviolet (UV) light. Trophoblast will bind to this fluorescent-tagged HLA-G antibody while other cells will not. When passed through the FACS machine, trophoblast cells will be picked up by a beam of UV light and selected into a special container. In this way, very pure preparations of trophoblast cells can be obtained. These cells are invaluable for research because they can be used for experiments in the secure knowledge that any results obtained will be derived from trophoblast and will not be due to the presence of contaminating cells.

Besides research, there are also potential clinical applications. Attempts have been made to use the antibody to HLA-G to isolate trophoblast cells circulating in the blood of pregnant women for prenatal diagnosis of genetic abnormalities in the fetus, such as Down's syndrome. For this procedure, cells of fetal origin are required for analysis (see Chapter 8). HLA-G has also been used to gauge the viability of embryos. Tests are now commercially available based on detecting HLA-G present in the fluid in which embryos are cultured before transfer into a patient undergoing IVF. The amount of HLA-G detected will reflect the health of the placenta and by extension the viability of the pre-implanted embryo so the clinician is able to select the best ones for transfer.

We now come to HLA-C, the third HLA present on the surface of trophoblast after -E and -G. Here, we have a different story altogether. HLA-C is rather an enigmatic molecule in many ways.[128] There are relatively few HLA-C molecules on the surface of cells, about 10 per cent of the levels found for -A or -B. Because of this low surface density, HLA-C had, in the past, been regarded as less important than -A or -B. Nor is it present in all primates but is seen only in humans and chimpanzees with a variable appearance in orang-utans, so HLA-C is a relative latecomer in primate evolution. Why has it evolved? One reason might be its involvement in the deeply invasive type of placentation seen in humans and some higher primates (see Chapter 5).

Unlike -E and -G, trophoblast HLA-C is polymorphic, so every placenta will have a different HLA-C inherited from the father, making -C the only one of the three trophoblast HLA molecules that will transmit a 'non-self' paternal signal to the mother.[129] Although polymorphic, HLA-C can conveniently be divided into just two groups, C1 and C2, in the way it binds to its receptor. Receptors for HLA-C on NK cells belong to a family known as KIR (killer-cell immunoglobulin-like receptor).[130] Like HLA-C, KIR is present only in humans and higher primates so the two seem to have evolved around the same time. KIR can also be divided into two groups, A and B. The A KIR group consists mainly of inhibitory receptors, whereas B KIR have activating as well as inhibitory receptors. Each pregnancy, then, is distinctive in having its own combination of trophoblast HLA-C (either C1 or C2) and maternal NK cell KIR (either group A or B). Some combinations are less favourable than others. As first demonstrated by Ashley Moffett's research group in Cambridge, the frequency of pregnancy disorders is increased when the maternal KIR is of the A type combined with a placenta which is HLA-C2.[131] An A KIR mother married to a C2 father is, therefore, bad for reproduction. Perhaps couples should be screened before marriage to avoid this association. C2 males should best be avoided, especially by A KIR women. Since A KIR consists mainly of inhibitory receptors, it seems that too much inhibition and/or insufficient activation is detrimental to pregnancy and will lead to disorders like miscarriage, growth retardation, and pre-eclampsia (see Chapter 5).[132]

Among populations worldwide, there is an inverse correlation between the frequency of HLA-C2 and A KIR. Those populations with a high frequency of A KIR have a low frequency of HLA-C2, whereas those which have a high frequency of HLA-C2 tend to be associated with the B KIR. This reciprocal pattern has emerged presumably because it reduces the chance of A KIR mothers carrying HLA-C2 babies and hence the risk of pathological consequences. This is a striking example of how pregnancy can influence the distribution of genes within a population. It has been known for some time that certain HLA-C/KIR combinations are advantageous in the body's defence against diseases, such as hepatitis C or AIDS. This is why these genes are so varied in different populations. Now we see that reproduction also provides a strong selective pressure for this variation.[133] But the combination required to deal effectively with diseases might not be the same as that which leads to pregnancy success. There is necessarily a compromise, which serves to explain the continual presence of gene combinations such as HLA-C2/A KIR that are unfavourable to pregnancy but presumably confer an advantage in defence against disease. Good for infection but bad for reproduction. This kind of trade-off or 'balanced selection' occurs frequently in the biological world.

The ability to deal efficiently with disease will be no use if the individual cannot breed successfully. It is no coincidence that the two systems are intimately linked. Immune response genes and genes for reproduction are the two most rapidly changing gene systems in mammals. These two systems probably evolved in tandem in order for placental viviparity to develop successfully. A study of the wild population of Soay sheep in a remote Scottish island by scientists from Edinburgh, confirms the relationship between reproduction and defence against infection. The island of Hirta in St Kilda is an inhospitable place. The sheep that live longest are those that have a strong immune defence system making them relatively less susceptible to disease. But there is a cost. They have fewer offspring compared with sheep whose immune system is less robust and which have a shorter life span. A strong immune defence against disease is acquired at the expense of fertility. Over the course of a lifetime, however, both groups tend to accumulate the same number of off-spring, ensuring that both genetic types of sheep continue to exist on the island.

From the preceding discussion, it should now be apparent to the reader that a certain degree of activation of the mother's immune system is required for successful pregnancy. This is somewhat surprising if we follow the organ transplant analogy. We would have expected there to be inhibition or, at least, avoidance of maternal immune response by the placenta. Instead, we find the opposite to be the case. So what is happening?[134] When uterine NK cells are activated, they do not necessarily start to reject and kill trophoblast as activated T cells would do to a transplant. Instead, these activated uterine NK cells might produce a variety of substances that actually encourage trophoblast migration into the uterus, a process which is essential for the modification of decidual arteries (see Chapter 5). This would explain why invading trophoblast populations have a polymorphic HLA-C on their surface rather than no HLA at all like syncytiotrophoblast. Signals from the father are required. The mother 'sees' this HLA-C from the father and will react to it. But this reaction is not necessarily to reject the placenta. Instead, this recognition of the paternal HLA-C on trophoblast will trigger the correct degree of activation of the mother's uterine NK cells, which is beneficial. In this way, she selects the paternal genes that best complement her own.

Attempts have been made to decrease the activity of NK cells to 'treat' women who are prone to suffering from miscarriage, the rationale being to stop these cells from attacking trophoblast. From what we have described, we can see that such 'treatment' has no scientific basis. Indeed, it could make matters worse since some NK cell activation is required for successful pregnancy. Some of the procedures are grotesque, such as injecting women intravenously with chicken egg yolk proteins. The fatty acids from the eggs are supposed to reduce the number of NK cells. Another frequently used regime to suppress NK cell functions is to give pregnant mothers large doses of steroids, such as prednisolone. This drug is largely metabolized by the placenta so little crosses to affect the fetus directly. But we should not be too complacent because it has now been shown that steroids will adversely impact on placental development which, in turn, can lead to inefficient transport of nutrients to the fetus. In an earlier chapter on fetal programming, we described how fetuses deprived of adequate nutrition in the womb are more prone to develop cardiovascular and metabolic diseases when they reach adulthood. So, while this regime of administering steroids may seem to confer immediate benefits in helping women with recurrent miscarriage to have a baby (this claim remains doubtful), the long-term cost is that these babies might not be too healthy when they grow up. Women desperate to have a baby now will probably not be too concerned about such future consequences, but they should be reminded of it.

Until we understand completely the immunological mechanisms underlying pregnancy, it is best to avoid any form of immune intervention to 'treat' miscarriages. Unfortunately, regimes of this kind are frequently offered to the public in spite of being scientifically unsound and not having been subjected to rigorous randomized clinical trials. The possibility of a 'placebo' (dummy) effect has not been considered. Miscarriage is a highly emotive condition and both patients and doctors are willing to try anything. One to 2 per cent of women will experience recurrent miscarriages (defined as three or more consecutive losses) in their reproductive lifetime. Pregnancy is particularly susceptible to the placebo effect because tender loving care is such an important psychological component, which, in itself, has considerable influence on the success or otherwise of pregnancy.

Since certain paternal HLA-C/maternal KIR combinations are unfavourable for reproduction, it would be a huge advantage if couples were able to select a compatible mate before pregnancy occurs rather than waiting to see if any pathological consequence shows up later. Mice can smell and discriminate between different H-2 molecules (the mouse equivalent of human HLA) excreted in the urine. Mice use these olfactory cues to choose a mate. We would expect them to select one whose H-2 is the same as their own in order to avoid potential rejection. But they don't. Instead, they opt for a mate whose H-2 are different and tend to shun those whose H-2 are too similar. This provides an opportunity for the pregnant female to react against the male's dissimilar H-2 in the same way as a mother recognizes the paternal HLA-C in human pregnancy. So here again we see that some maternal response to paternal signals is beneficial to pregnancy. Are humans able to use smell for mate choice? Early mammals mostly used their olfactory senses because they were mainly nocturnal animals, foraging for their food on the forest floor. Many modern animals continue to have an acute sense of smell and use it for their daily activities. Their nasal cavity is large and the area is further increased by bony ridges extending into it from the roof and sides. A dog's sense of smell, for example, is many thousand times more acute than a human's. Their noses have 300 million olfactory receptors whereas we have only about 6 million. Also, the olfactory region of the dog's brain for processing smell is considerably larger than in the human brain, so they can make more sense of the scents they have inhaled than we can. There is also an additional structure in many animals known as the 'vomero-nasal organ', situated in the floor of the nasal cavity, which is used primarily to detect 'pheromones'. These are agents involved in social and sexual signalling between animals.

In humans, olfactory senses are supposed to have been largely superseded by audio-visual perception as we evolved from a nocturnal to a diurnal animal. Olfactory cues no longer have the kind of influence on our behaviour that they have for lower animals. Darwin himself felt that the use of smell might have been important to our ancestors but plays a less significant role in modern man. But is this true? Is our olfactory ability really inferior to that of other mammalian species? How do we rank in the olfactory perception scale? Smell scientists are beginning to argue that we humans are better at smelling than we give ourselves credit for. The book by Avery Gilbert with the evocative title *What the Nose Knows* is full of entertaining information. The difference between humans and animals is not so much the ability to detect odours, but how the brain interprets the olfactory signals. For many animal species, smell triggers an inbuilt reflex response, for example, to flee immediately when the smell of a predator is detected or to choose a mate if there is a good match in odour. The response is involuntary.

In contrast, humans have higher cognitive abilities which allow us to decide how to use the original odour signal. Even if we can detect a suitable HLA match in a potential mate, we may ignore the signal and go for other criteria, such as beauty or intelligence. We have a choice. Have humans retained the ability to discriminate between different HLA by smell as mice can for H-2 in mate choice? To distinguish HLA-C1 from C2 might be too much to ask, but might some broad distinction be possible? There have been experiments conducted in which college students were handed worn sweatshirts from students of the opposite sex and then asked which student they found to be the most attractive, based on the smell. Some studies have reported that, like mice, students also showed a preference for someone with disparate HLA rather than one who's HLA was similar. The results are equivocal because of the small sample sizes. So these observations remain anecdotal and unproven.

Furthermore, the advent of external agents such as perfumes is likely to mask any natural olfactory cues present. Women on the contraceptive pill are said to have different preferences for potential mates from those not on the pill, using similar 'sweatshirt' experiments as described earlier. Presumably this can be put down to the effects of hormonal changes on olfactory perception. A woman might then select a certain type of mate while on the pill, but when she wants to start a family and comes off the pill after marriage, she suddenly finds that she has made the wrong choice and is no longer attracted to her husband. Could this be a cause of marriage breakdown?

We are all familiar with the experience of a particular human odour being more acceptable or even more sexually attractive than another, but whether this preference has an immunological basis is not clear. According to Mairi Macleod, writing in *New Scientist*, it was a common practice for maidens in Elizabethan England to place slices of apple in their armpits to absorb the smell and then present these to potential suitors. And in morris dancing, men put handkerchiefs in their armpits before waving them under the noses of young females. These are both symbolic gestures demonstrating the allure of personal odour as a potent force for sexual attraction. The composition of this 'eau de armpit', as *New Scientist* drolly dubbed it, deserves further investigation. The fact that such rituals exist means there must be some truth behind it. Are HLA molecules involved in determining personal odour and could these molecules lead to subliminal olfactory effects in human mate choice?

Reproduction and transplantation, then, have different requirements. For successful organ transplantation, the aim is to 'match' the HLA of donor and recipient as closely as possible to reduce the risk of rejection, whereas in reproduction some maternal recognition of paternal HLA seems to be preferred.[135] This realization has completely transformed the way in which we think about the immunology of the placenta. The whole paradigm has shifted. In the past, most research efforts were focused on how the placenta might evade the mother's immune response. Innumerable hypotheses were proposed, none of which withstood the test of time. This is not surprising. With hindsight, we now know that we have been following a false trail. The placenta, it turns out, is not like an organ graft after all, in spite of superficial resemblance. The placenta does not have to duck under the radar of the mother's immune system and avoid its surveillance. Instead, a certain level of maternal recognition and response is essential for pregnancy to be successful. After all, the reason for having sexual reproduction is to broaden the genetic variation within the population. The mother is encouraged to have children whose HLA is different to herself as this will improve resistance to disease. For this, evolution has adapted the mother's immune surveillance to view the placenta in her womb, not as an alien intruder that should be thrown out, but as a close relative who is welcomed to stay for a while, if she considers this visitor worthy of her hospitality.

7

Products

In addition to all the other jobs it performs, the placenta is also a factory busily manufacturing a wide range of products to sustain pregnancy. The best known are the placental hormones. Many of these will replace those made by the mother herself so that the placenta gradually takes over the reins from the mother and plays an increasingly dominant role throughout pregnancy. The mother is relegated to the back seat.

Hormones are normally made in endocrine glands.[136] We have two kinds of glands in our body: exocrine and endocrine. Exocrine glands secrete substances into ducts that empty into body cavities or outer surfaces. Examples are sweat glands of the skin, salivary glands, and digestive glands of the stomach and intestines. Endocrine glands do not secrete substances into ducts which is why they are sometimes called 'ductless glands'. Instead, their secretory products, hormones (Greek 'hormon' = setting in motion), are released into the spaces surrounding the endocrine cells from where they enter the bloodstream and are carried to other parts of the body where they exert their effects. Hormones, then, are chemical messengers which influence the activity of other tissues of the body even when those tissues are at a distance from the glandular source of production. Although the whole body will be exposed to these hormones since they are transported via the circulation, only certain 'target' tissues are responsive because these have specific receptors that can bind and react with a given hormone. Those tissues without the appropriate receptors will be unaffected.

Classical examples of endocrine glands are the hypothalamus, pituitary (anterior and posterior), pineal, thyroid, parathyroid, adrenals (cortex and medulla), pancreas, and gonads (testes and ovaries). The placenta is not normally included among this group in endocrinology textbooks, but it should be. The placenta secretes hormones vital to pregnancy.[137] Most of these are not intended for the fetus. They are predominantly for maternal consumption. This was recognized over 100 years ago when a number of 'placental products' (they were not known as hormones then) were extracted from the placenta and observed to have similar effects on the maternal uterus and breast as those from her own ovaries. All these glands comprise the 'endocrine system', a term first coined by the physiologist Ernest Starling in 1905. The functions of hormones are to regulate homeostasis, that is the maintenance of the body's internal systems, such as blood pressure and temperature at a certain constant level despite variations in the external environment, as well as metabolism and reproduction. The endocrine system is associated with the nervous system and the functions of the two are closely interrelated.

Structurally, hormones belong to two major families: peptide hormones and steroid hormones. Peptide hormones are more abundant. They are made up of chains of amino-acids which are the building blocks for proteins. Some are very small with only a few amino-acids while others consist of long chains. Peptide hormones are stored inside the endocrine cell ready made in the form of 'secretory granules' for immediate release by a mechanism known as 'exocytosis' in response to appropriate signals. Steroid hormones, in contrast, are made from cholesterol. To most people, the mention of cholesterol is inevitably associated with coronary heart disease but cholesterol is a vital ingredient for the synthesis of steroid hormones. They have a common core structure and small chemical variations in this core give rise to the different hormones. Unlike peptide hormones, steroid hormones are not stored in the endocrine cell as secretory granules. Instead the cell keeps a store of cholesterol rather than the hormone itself. This is because steroid hormones are very soluble in the fatty environment within the cell and are difficult to store whereas cholesterol can be easily stored and then synthesized rapidly into the relevant hormone when required. Because it is fat soluble, a steroid hormone, once made, just diffuses across the cell membrane into the exterior without the need for a specialized secretory mechanism such as exocytosis used by peptide hormones.

In the bloodstream, hormones circulate in very low concentrations so that they are measured in extremely small units such as nano-moles per litre (nmol/L) while most other substances exist in micro-moles per litre quantities or greater. Steroid hormones do not circulate freely in the blood but are bound to what are known as 'carrier proteins'. These protect the hormone from being degraded and thus prolong the life of the hormone in the circulation. The carrier proteins also provide a readily accessible reserve of the hormone in the blood because only the unbound fraction of the hormone is biologically active. Peptide hormones are rapidly inactivated by digestive juices in the stomach while steroid hormones are not. This is why the latter can be given by mouth when required therapeutically but the former, such as insulin, have to be injected.

The specific receptors for peptide hormones are usually found on the outer surface of the target cell because these hormones cannot get inside the cell. In contrast, since they are fat soluble, steroid hormones can easily pass through the cell membrane of the target cell, so receptors for steroid hormones are usually found on structures inside the cell such as the nucleus. After interacting with appropriate receptors, hormones bind to certain areas of genes which will alter the behaviour and activity of the target cell. This is how hormones work. They can act on tissues remote from the hormone-producing cell (endocrine action), on tissues nearby (paracrine action), or even on the hormone-producing cell itself (autocrine action). Many hormones function in a 'cascade' fashion. For example, a hormone A from the hypothalamus can stimulate the release of another hormone B from the pituitary which in turn stimulates the release of a hormone C from the thyroid gland. This is called the 'hypothalamus–pituitary–thyroid axis'. There are many advantages of such a system, which include flexibility, amplification, and fine tuning. There is also the phenomenon of 'negative feedback' when the final product inhibits the release of hormones higher up the cascade, thereby shutting off further production when an optimum supply is reached.

The human placenta produces both peptide and steroid hormones. The peptide hormones are human chorionic gonadotrophin (HCG), human placental lactogen (HPL), and corticotrophin-releasing hormone (CRH). The steroid hormones are progesterone and estrogen. Many of these hormones are structurally and functionally very similar to those secreted by the mother's pituitary, hypothalamus, and gonads. Because of their similarity in structure, placental hormones can mimic the functions of hormones from these maternal organs and gradually take over from the mother. The changes in levels of two of these hormones throughout pregnancy are illustrated in Figure 14.

The earliest hormone made by the human placenta is human chorionic gonadotrophin (HCG). It is synthesized by trophoblast as early as 48 hours after implantation so it is the first sign the mother has that she is pregnant. This is referred to as 'biochemical pregnancy'

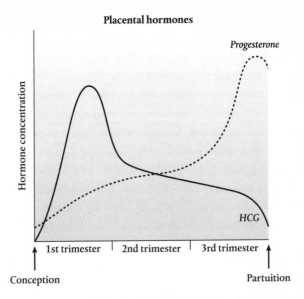

FIGURE 14 Graph showing the levels of HCG and progesterone throughout pregnancy.

and precedes 'clinical pregnancy' which is based on the detection of
the embryo itself by clinical means such as ultrasound. The earliest
that HCG can be detected is at around 5 weeks of pregnancy. The
main source of production is the syncytiotrophoblast.[138] Since syncytiotrophoblast is in direct contact with the mother's blood, the effects
of HCG are mainly directed at the mother. The levels of HCG in pregnant women's blood rise steadily, reaching a peak in the second month
of pregnancy, and then decline (see Figure 14). Chorionic gonado-
trophins are synthesized in the placenta of most, if not all, eutherian
mammals. The horse, for example, has high levels of equine chorioic
gonadotrophin (ECG), although production is relatively low in the
placentas of sheep, cows, and pigs compared to humans and other
primates. While the synthesis of HCG begins just after implantation
in humans, this synthesis can precede implantation by several days

in the pig, sheep, and cow. The synthesis of chorionic gonadotrophin by the placenta, therefore, is a very early event in mammalian pregnancy.

Because of its early production, the presence of HCG in a woman's blood or urine is used as a test for whether she is pregnant. In the old days, the test used was a bio-assay, so called because the patient's blood or urine was injected into animals, such as mice or toads. Immature female mice injected with the urine of pregnant women were killed a few days later and their ovaries and uteri examined. If HCG was present in the urine, these organs would be enlarged. Mice were subsequently replaced by the female Xenopus toad. When injected with urine containing HCG, the toad would ovulate within 12 hours. The advantages of using toads over mice were two-fold. The toads did not need to be killed so were reusable. Also, the results of the test were available the next day rather than several days later. But the major inconvenience of bio-assays is that their sensitivity is such that HCG is not usually detected until 2–4 weeks after the last menstrual period. This makes diagnosis of pregnancy a bit late in the day for those women who, for whatever reason, need to know the result quickly.

The modern technique is an immuno-assay, using a specially designed antibody which recognizes HCG. Most of the 'home test' kits for pregnancy are based on this method. If present, HCG will bind to the antibody, resulting in a change in colour of the fluid in the kit. Immuno-assay is an improvement over bio-assay in that the test is easy to use and results can be obtained within a few minutes. A chart normally accompanies a test kit illustrating the amount of HCG likely to be detected if the woman is pregnant, depending on the stage of gestation. A more sophisticated variation of immuno-assay is radio immuno-assay using a radioactive material in combination with an antibody to monitor the presence of HCG. This is an extremely sensitive technique and can detect even minute traces of HCG but, unfortunately, specialist equipment is needed which precludes its use in the home. A major drawback of immuno-assays is the occurrence of false positives and false negatives. If the antibody does not have a tight enough fit for HCG, it might react against some other substances as well as HCG. This is known as 'cross-reactivity' and will lead to false positive results because the presence of another substance is mistaken for HCG. Women who do not wish to be pregnant will be unnecessarily alarmed. The corollary to this is that the antibody is not specific enough for HCG and, therefore, does not recognize the hormone even if it is present. This will lead to a false negative result. Another kind of false negative is not the fault of the technique itself but due to the actual absence of HCG in the blood or urine even when the woman is pregnant. Such a situation can occur in ectopic or tubal pregnancy. This can lead to misdiagnosis of an ectopic pregnancy with potentially fatal consequences. The absence of detectable HCG should not negate the clinical suspicion of an ectopic pregnancy.

Reproduction in the human female is centred on the menstrual cycle. The average cycle lasts for 28 days, hence the term 'menstrual' meaning 'monthly', but can vary between 21–35 days. Conventionally the days of the menstrual cycle are numbered from day 1 which is the first day when bleeding starts and a woman experiences her 'period'. This usually lasts for 4–7 days and might be accompanied by severe abdominal cramps known as 'dysmenorrhoea'. After shedding, the uterine endometrial lining regrows, the initial stages (from days 7–14) being labelled as the 'proliferative' phase leading into the 'secretory' phase (from days 14–28). Ovulation (release of an egg) usually occurs at mid-cycle around day 14. All these events are the result of two gonadotrophic hormones secreted by the anterior pituitary: follicle stimulating hormone (FSH) and luteinizing hormone (LH). A gonadotrophin is a hormone whose activity is directed at the gonads which, in women, are the ovaries. A sudden increase in the production of LH by the pituitary is the stimulus for ovulation. This so-called LH 'surge' immediately precedes ovulation and can be used to time when an egg is about to be released. In addition to the menstrual cycle, many biological phenomena have natural rhythms of which the most familiar example is our heart beat. There is also the circadian rhythm which determines our response to the cycle of day and night.

Following ovulation a corpus luteum (yellow body) is formed from the wall of the follicle that produced the egg. The major steroid hormone produced by the corpus luteum is progesterone, which causes dramatic changes to the lining of the uterus, preparing it for pregnancy. This is the process of 'decidualization' without which the placenta will not be able to implant (see Chapter 5). Progesterone is 'thermogenic', meaning that it raises the body temperature. A woman's fertile period each month can be identified by observing the small rise in her body temperature caused by the increased progesterone secretion following ovulation. The corpus luteum has a limited life span of around 14 days. In the absence of a fertilized egg, the corpus luteum starts to degenerate and the level of progesterone it produces begins to decline. Without the continual stimulus of progesterone the uterine lining will break down, leading to the onset of menstruation. In Chapter 6, we described the part played by NK cells in initiating the onset of menstruation. The viability of NK cells is dependent on progesterone without which they will not survive, so the action of progesterone on the endometrium could be through its effects on NK cells.

If pregnancy occurs, the corpus luteum does not degenerate but continues to secrete progesterone during the first 3 months of pregnancy. Progesterone then, is a vital hormone that dominates female reproduction, from producing dramatic changes in the uterine lining during the latter part of the non-pregnant menstrual cycle to maintaining decidua once pregnancy has occurred. The artificial induction of labour by the compound RU486, which is a progesterone antagonist, testifies to the importance of this hormone in sustaining pregnancy. The crucial question then is what extends the life of the corpus luteum in pregnant women and makes it continue to produce progesterone? HCG secreted by trophoblast is responsible for this. By this ingenious manoeuvre, the placenta successfully manipulates the mother's ovarian function to ensure its own survival. We described earlier how LH, a gonadotrophin produced by the mother's pituitary, is involved in ovulation and the subsequent development of the corpus luteum. HCG is also a gonadotrophin capable of acting on the ovaries. HCG and LH are descended from the same ancestral molecule during evolution and share a common structure (see later). HCG can take over the role of LH to prolong the life of the corpus luteum, a case of placental gonadotrophin replacing the activity of a maternal pituitary gonadotrophin. The placenta is behaving like the mother's brain in directing ovarian function but the stimulating placental gonadotrophin and the responding maternal ovary are from two different individuals, a situation that is unique in biology.

The continual secretion of progesterone by the corpus luteum helps to support the early development of the placenta. But by about the 5th week of pregnancy, the placenta itself begins to produce its own progesterone and also some estrogens. The levels of these hormones produced by the placenta increase throughout pregnancy (see Figure 14). Estrogen was first extracted from the human ovary in 1923 and from the placenta in 1927, while progesterone was extracted from the corpus luteum in 1929 and from the placenta in 1932. Placental synthesis of these hormones is also stimulated by HCG, another important role for this gonadotrophin. The placenta has now taken over the function of the mother's corpus luteum in maintaining its own survival. This is why removal of the corpus luteum before the 7th week of pregnancy causes miscarriage but does not after this time, because the placenta has made the corpus luteum redundant. The placenta is effectively saying to the mother, 'I don't need you anymore. From now on, I can take care of myself and the baby'. The sheep placenta also produces progesterone. As in humans, the pregnant ewe's ovaries can be removed after mid-gestation without disturbing the pregnancy because the placenta has taken over. In contrast, the goat placenta does not produce progesterone so pregnancy is dependent on secretions from the maternal ovaries throughout. This is surprising because goats and sheep are closely related. What happened during the lifetime of their common ancestor that led to these two species doing things so differently today?

The corpus luteum might have had an entirely different function early in its evolutionary history. In marsupials, putting a newborn onto the nipples of a virgin female will stimulate her mammary glands to lactate. The female does not have to be pregnant. The breast is already primed to produce milk and the stimuli for this are hormones released by the corpus luteum formed after each ovulation. The earliest mammalian corpus luteum, therefore, may have been concerned mainly with the development of mammary glands as seen in marsupials and it was only later that it directed its attention to sustaining the placenta in eutherians, demonstrating once again that lactation preceded placentation in mammalian evolution.

Prolongation of the life of the corpus luteum by HCG is often referred to as 'maternal recognition of pregnancy'. This is the earliest time when a chemical signal (HCG) from the embryo is recognized by the mother. The crucial importance of HCG during the early stages of pregnancy should now be apparent. The maintenance of the pregnancy depends on the continual secretion of progesterone by the corpus luteum and later by the placenta itself. Both these events are dependent on HCG. Many people are surprised that post-menopausal women even at age 60 or beyond can become pregnant if they are provided with eggs from a donor. This is because once the donated egg is fertilized and implanted, the placenta can take over all subsequent endocrine control necessary to sustain pregnancy. The loss of reproductive ability with age in women is primarily due to the absence of healthy eggs and not because of senescent changes in the uterus. The infrastructure, as it were, for sustaining pregnancy continues to be present for a considerable time after the menopause.

Women are now starting families at an older age.[139] This is when the decline in both quantity and quality of eggs in the ovary occurs. At four months in the womb, a female fetus has about 6–7 million egg follicles in the ovary, but this number is reduced to 1–2 million at birth, so there is already a substantial loss during intrauterine life. At puberty, around 300,000–400,000 follicles remain with a final decrease to less than 1000 by the time of the menopause. Accompanying this decline in number is a reduction in egg quality that will lead to an increasing incidence of embryos with chromosomal defects conceived by ageing mothers.

Since the fundamental problem is inherent in her own ageing egg, IVF alone cannot help an older woman to conceive a healthy baby. One solution is to fertilize an egg donated from a younger woman and implant this into the ageing mother. This technique has prolonged the fertile period of women significantly, with the oldest reported success being in a woman of 70 (although the record might be said to be held by Abraham's wife, Sarah, who according to the Bible became pregnant at the age of 90). But egg donation is not entirely free from complications. The resultant placenta will have no genes in common with the recipient mother, unlike in a normal pregnancy in which half the genes in the placenta are inherited from herself with the other half provided by the father. This, then, is an example of 'missing-self', a phenomenon we discussed at length in Chapter 6. Pregnancies resulting from egg donation have a higher incidence of pre-eclampsia because the absence of her own genes will trigger a reaction from her innate immune response. An alternative approach would be for a woman to freeze her own eggs when she is young for use in the future. Some women are already doing this. It will avoid the immunological problems associated with donated eggs. To date, there is still only limited experience of this technique so it is too early to evaluate the safety of oocyte cryo-preservation. How long can eggs be kept frozen without deteriora-tion? At present, we don't know.

A new and exciting era is about to emerge. Progenitor cells capa-ble of developing into immature eggs or oocytes have been found in the human ovary, so women are not born with a finite number of eggs as was once thought. Hidden within her ovaries is a reservoir of potential egg-producing cells that can be brought into play when needed. A woman's fertility should no longer be constrained by the depletion of eggs as she ages. Her egg pool is constantly renewable throughout life in the same way as a man's testicular generation of sperm. Gender equality in gamete production is here.

The placenta makes the steroid hormone progesterone by directly converting cholesterol from the mother's blood, but to make estro-gen the placenta needs the help of the fetus. Progesterone made by the placenta passes to the fetus and, in the fetal adrenal glands, is converted into a variety of substances. These are then carried back to the placenta via the fetal blood stream and are then converted into estrogen. It is interesting that the ovaries, which also synthesize estrogens, already contain all the enzymes involved in the conver-sion of cholesterol. By contrast, some of these key enzymes are missing in the placenta. While the placenta is fully equipped for active conversion of cholesterol from the maternal blood into pro-gesterone, it cannot by itself make estrogen. It has to rely on col-laboration with the fetus for this task. This cooperation between the fetus and the placenta has given rise to the term 'fetoplacental unit' to emphasize the close association between the two. Increased levels of estrogen can give rise to skin rashes or red blotches in pregnant women. Although somewhat unsightly, they are, nevertheless, de-scribed ardently as the 'glow' of pregnancy. In contrast to progester-one, the action of placental estrogen is directed mainly towards the fetus in helping the development of its organs, like the lungs, kidneys, and liver. Placental progesterone, on the other hand, is more for maintaining pregnancy, so the two hormones have different roles to play. The placenta makes products to look after itself as well as the fetus.

The mechanism of production of HCG is unusual.[140] Most peptide hormones, as mentioned earlier, are concentrated and stored inside the cell that produces them in structures known as 'secretory granules'. They are like little balloons stuffed with hormones. At the right moment they burst, releasing the hormones inside them into the exterior by the process of exocytosis and these are then carried throughout the body by the blood stream. Secretion of HCG from trophoblast is different. It is not packaged in secretory granules. Instead, it makes its way to the surface of the cell from where it breaks off and is then swept away into the circulation. This results in two forms of HCG. One is the secreted form with hormonal functions. The other is that which is left behind on the cell surface.[141] What might be the function of this surface HCG? One theory is that this HCG protects the surface of syncytiotrophoblast from immune attack by the mother. The HCG molecule is a sialoglycoprotein, that is, it is rich in sialic acid. Sialic acid possesses a strong negative charge which makes it repel other molecules. The non-clotting, Teflon-like surface of syncytiotrophoblast described earlier in Chapter 3 could be contributed to by surface HCG. This negative charge might also keep maternal immune cells at a safe distance. Here again we see a molecule with a dual function. We have encountered similar situations in preceding chapters where Nature has economized.

A novel form of HCG has been described called HCG-H, the suffix 'H' standing for 'hyperglycosylated'.[142] This means a larger saccharide (sugar) component making up the HCG molecule. The evolution of the saccharide residues in the CG molecule shows an interesting trend. In earlier primates (e.g. Cebus monkey), there are only two saccharide residues in their CG molecule, the number increasing to three in later primates (e.g. orang-utan), and then to four in the Great Apes and humans. This increasing number of saccharide residues prolongs the life of the molecule in the circulation thereby enhancing its biological activity. In humans, the four saccharide residues in regular HCG are all trisaccharides whereas the saccharides in HCG-H are hexasaccharides, which explains why the latter molecule is approximately twice as large.

HCG-H is said to be the predominant form present in early pregnancy, comprising over 90 per cent of total HCG in the 3rd week of pregnancy. This proportion then declines in subsequent weeks until it makes up less than 2 per cent of total HCG by the 20–30th weeks of pregnancy. While regular HCG is synthesized by the syncytiotrophoblast as described earlier, HCG-H is produced by cytotrophoblast. Both types of HCG have a significant influence on implantation, with regular HCG promoting blood vessel development in the uterus and HCG-H enhancing trophoblast growth and invasion. Diseases such as pre-eclampsia, which is the result of inadequate trophoblast invasion, could then be associated with low levels of HCG-H production. This is the reason for obstetricians' interest in this molecule, because measuring its level in a pregnant woman could identify patients who are at risk. At present, there are no reliable predictive markers for pre-eclampsia. It must be emphasized that all the above proposals are presently just theories but they are worth considering. The only proven function of HCG so far is its action in prolonging the life of the corpus luteum.

Choriocarcinoma (CHO), a cancer derived from placental trophoblast (see Chapter 3), secretes vast quantities of HCG.[143] This is to be expected since many cancers continue to function like their normal counterpart. Estimation of HCG in the blood of patients with CHO is a routine way to monitor the progress of the disease and to see if there is any cancer left after treatment. Even minute residual deposits of cancer, which might be too small to detect otherwise, will leave a conspicuous trail of HCG. What is surprising, however, is the observation that many other cancers besides CHO, even those of non-trophoblast origin, frequently produce HCG. This is known as 'ectopic' production.[144] For example, some stomach cancers can secrete amounts of HCG into the blood of the patient approaching the peak levels found in the early stages of pregnancy. Stomach cells do not normally produce HCG. This is a further example of the sharing of characteristics between cancer and trophoblast. Indeed, it is claimed that if we were to look for the presence of HCG on the surface of cancer cells (membrane form) rather than in the blood of the patient (secreted form), then all types of cancers, regardless of tissue origin, would show a positive reaction. What this means is that the genes responsible for synthesis of placental HCG, which are normally switched off in other mature adult cells except for trophoblast, are reactivated when these cells become cancerous.[145] Surface HCG could play a similar role in cancer cell survival as it is said to do for normal trophoblast.

The list of similarities between trophoblast and cancer is impressive. In their behaviour, both are highly invasive and have the ability to destroy blood vessels. They synthesize similar products, such as HCG, which is referred to as an 'onco-placental' protein, the prefix 'onco' meaning associated with cancer. They share the same cell surface antigenic characteristics, such as low levels of HLA molecules. The immune response of the mother to her placenta and the host response to cancer both utilize NK cells and innate immunity. A phrase frequently used by cancer biologists is that 'neoplastic transformation is accompanied by regression towards an embryonic phenotype'. In other words, the nature of cancer is when a mature adult cell starts to behave childishly. This has led to attempts to treat cancer by administering 'differentiating agents' to persuade cancer cells to revert to their normal adult behaviour instead of killing them by radiotherapy and/or chemotherapy. It is tempting to conclude that the mechanisms operating in cancer cell survival are the same as those utilized by trophoblast in normal reproduction. Cancer may have hijacked them for its own ends.

The importance of HCG, estrogen, and progesterone in maintaining pregnancy[146] led to the idea that inadequate levels of these hormones might be the cause of miscarriage in some women. Doctors have tried to build up these hormones in those women who show signs of impending miscarriage in the hope of rectifying this imbalance. So far, no definite benefits have been observed, but it is the potential harm of such regimes that must be borne in mind. This is seen in the disastrous story of the synthetic estrogen, diethylstilbestrol (DES). In some ways, the story of DES is even more tragic than that of thalidomide because it took so long for the danger to be recognized. Many clinicians were already wary about the use of a synthetic estrogen, such as DES, which was much more potent than the naturally occurring substance. Also, early clinical trials on DES were inconclusive and mostly showed that the drug was not particularly effective in preventing miscarriage. In spite of these misgivings, the pro-lobby was so persuasive that DES was approved for use in pregnancy in 1947. The crucial argument was that no harmful effects were observed in either the mothers or their babies. But these effects appeared dramatically 15–20 years later. In the late 1960s, a group of young women whose ages ranged from 15–22 were diagnosed with a rare form of cancer in the vagina called adenocarcinoma. Subsequently many more cases were identified. This type of cancer is normally uncommon in women of this age group. The alarm was triggered when a mother of one of the young women affected by vaginal cancer asked her doctor whether this could be linked to her taking DES when she was pregnant. Subsequent investigations confirmed that DES was indeed responsible. DES can cross the placenta, creating an abnormal hormonal environment in the womb during a critical period when the baby's organs are developing. This can lead to cancer in later life, a further example of fetal programming described in Chapter 5. A number of recent studies have raised the suspicion that women exposed to DES *in utero* might also have an increased risk of developing breast cancer. DES was finally banned in 1971 but, by then, millions of women had already taken the hormone. The dream of averting miscarriage had become a nightmare.

Reproductive biologists are constantly on the lookout for effective methods of fertility control to complement those which are currently available, such as the contraceptive pill. A vaccine that can prevent pregnancy in the same way as for infectious diseases would be an ideal solution.[147] One that has been tried is a vaccine against HCG. The reason why HCG has been chosen as a target is because it is secreted exclusively by trophoblast cells of the placenta for use in pregnancy so the vaccine will not disrupt other metabolic activities. The importance of HCG in the maintenance of pregnancy is well established as we have seen earlier, so its destruction by a vaccine would be expected to disrupt pregnancy. HCG would appear to be an appropriate target. Unfortunately, the practical problems associated with the production of an effective anti-HCG vaccine have proved more numerous than anticipated.

Although HCG is unique in the sense that it is produced exclusively by trophoblast, the structure of the molecule itself contains parts which are similar to other hormones so that potential cross-reactivity is a major hindrance. HCG belongs to a family of peptide hormones such as leuteinizing hormone (LH), follicle stimulating hormone (FSH), and thyroid stimulating hormone (TSH) produced by the pituitary whose structures are made up of two parts, known as α and β subunits.[148] The α subunit is common to all members of the family and therefore cannot be used to distinguish HCG from other members of the family. It is the β subunit which differs between members of the family and which determines the function of each hormone. Even within the β subunit there are large areas of the molecule which are shared. CG first evolved from LH by a single deletion mutation in the LH β subunit. Interestingly, this deletion is not present in the primitive prosimians (lemurs, lorises, tarsiers) but appeared only in the simian primates (monkeys, apes, humans) so

CG is a relatively recent acquisition. Subsequently more changes occurred in the CG β subunit among the simians, leading to their own CG such as HCG for humans. In the case of HCG there is a tiny part at the end of the β subunit that is unique and differentiates it from all other hormones of the same family. A vaccine must be directed at this part of the HCG β subunit if it were to be specific. Otherwise, the vaccine will have unwanted side effects on other hormones. Such a specific vaccine has been made.

Limited clinical trials have confirmed the safety of this vaccine but one major drawback is the wide variability in response encountered among individual patients.[149] As HCG is produced in large quantities during early pregnancy, it is to be expected that only those individuals capable of producing enough antibodies to neutralize all the hormone would be adequately protected against pregnancy. The vaccine has been tested on marmoset monkeys and it has been confirmed that, if there is a good immune response to HCG, then pregnancy is prevented. But if the response is inadequate, then the female monkeys experienced recurrent miscarriages interspersed with occasional live births. This vaccine, in its present state of development, obviously cannot be sanctioned for human use. Finally, an anti-HCG vaccine, even if successfully produced, is effectively an abortifacient rather than a true contraceptive in that it disrupts the maintenance of pregnancy *after* conception has taken place. This might not meet with the approval of large sections of the community either for ethical or religious reasons. All this is deeply disappointing because the concept of an anti-fertility vaccine is intellectually rather appealing. Still, it was worth trying.

We have already alluded to the vital role played by progesterone in maintaining pregnancy, but its action is not confined to this stage alone. Progesterone is directed at other maternal targets. Among these is the breast, where placental progesterone prepares the breast tissue in readiness for subsequent milk production. The other target is the mother's brain, where progesterone modifies her behaviour.

During pregnancy, high levels of progesterone reduce sexual activity in spite of the fact that, in humans and other higher primates, sexual receptivity has been substantially emancipated from hormonal influence. After birth, progesterone modifies maternal behaviour towards caring for the young offspring. This encourages maternal–fetal bonding (see Chapter 4). The 'motherly' mood some women experience during the latter part of the menstrual cycle is thought to be related to the high levels of progesterone circulating in her body at this time.

The human female is unique in having a prominent display of her breasts even when not lactating. She is the only primate to do so. The fatty tissue in her breast has become responsive to estrogen secreted by her ovaries and this accumulation of fat shapes the adult female breast. This is different to enlargement of the lactating breast which is due to proliferation of the milk-secreting glands stimulated by progesterone released by the placenta. The human female breast has assumed an additional function: besides being an organ for feeding the young, it has also become a sexual attractant. As we have discussed in Chapter 6, visual cues have largely replaced olfactory senses in human mate choice.

Another peptide hormone produced by the placenta is human placental lactogen (HPL).[150] This hormone is produced later in pregnancy than HCG. Together with estrogen, HPL helps progesterone to prepare the breast for milk production by stimulating proliferation of the milk-secreting glands in the breast. After birth, suckling by the baby stimulates the sensory nerve endings of the nipple, sending signals to the anterior and posterior pituitary glands at the base of the mother's brain. The anterior pituitary releases the hormone 'prolactin'. As its name implies, prolactin initiates milk production by the glands in the breast. At the same time, the posterior pituitary releases another hormone 'oxytocin' which causes constriction of the smooth muscles around the glands, to expel milk from the breast. Oxytocin is also known as the 'attachment hormone' and plays an important role in mother/infant bonding during breast feeding. Besides affecting the mother's brain, oxytocin also travels to the infant's brain via breast milk. It encourages calm, euphoria, and love in both individuals. A nasal spray containing oxytocin can be bought over the counter in pharmacies for anxious individuals! The veterinary profession is already aware of this. Lactating bitches naturally release a pheromone which makes puppies secure and relaxed during the postnatal period. A dog collar impregnated with a synthetic compound mimicking the activity of this pheromone is available. I have one of these collars for my rather nervous Labrador.

Perhaps a more important function of HPL is that it mobilizes fat from the mother's deposits to be used for her own fuel requirements. She will now use fat rather than glucose for her metabolism. In this way it causes the mother's blood sugar to rise, providing the fetus with abundant glucose for its growth. Structurally, HPL bears a remarkable resemblance to human growth hormone (HGH) produced by the mother's pituitary gland. Approximately 80 per cent of the building blocks that make up the two hormones are identical. This suggests that the two hormones may have evolved from the same ancestral molecule. We have two hormones derived from different sources (placenta and pituitary) having the same result (promoting fetal growth) but in different ways. All this is reminiscent of the close evolutionary, structural, and functional relationship between placental HCG and pituitary LH described earlier. The fetal placenta and the maternal brain do seem to have a great deal in common.

The signals that initiate labour in humans have not been clearly established. The third peptide hormone produced by the placenta, corticotrophin releasing hormone (CRH), could influence this process. Its secretion comes very late in pregnancy, approximately 20 days before the start of labour (parturition), so its appearance is timely for this role.

The importance of placental endocrine activity should now be clear. Its hormones are crucial for the establishment of pregnancy, for its maintenance throughout gestation, play a role in the induction of labour, and affect both the mother and infant after birth. This is a remarkable achievement for one organ.

Unlike other hormones whose functions are to transmit information to different parts of our own body, placental hormones are produced by one individual (fetus) to influence the activities of another individual (mother). The sender and receiver of placental hormones is no longer the same person. This adds a new dimension to placental hormonal action. As we have already noted, placental hormones are identical or at least very similar in structure to the mother's own hormones which she produces when she is not pregnant. The receptors on her tissues which are targets for these hormones cannot distinguish whether the source of the hormones is herself or the placenta. She is tricked into responding. The mother has now relinquished control of her own bodily functions to signals coming from the placenta. The fetus has taken over. This is a commendable act of self-sacrifice by the mother to ensure the success of her pregnancy. In Chapter 5, on implantation, we already encountered another example of maternal concession. She allows placental trophoblast cells to tap into her own circulation to access nutrients from her blood to feed the baby.

Another puzzling thing about placental hormones is that they are produced in massive amounts. Their output is far in excess of requirement. They seem not to be subjected to the same kind of checks and balances seen with other hormones, such as the negative feedback loop described earlier which serves to control redundant production. Why is this? David Haig has a provocative idea.[151] He likens it to an evolutionary arms race between the fetus and mother in which placental hormones are manufactured in ever greater quantities to elicit favourable responses from the mother for the benefit of the fetus, while the mother evolves to be less and less responsive in order to defend herself against immoderate fetal demands. This brings us back to the fetal–maternal 'conflict' hypothesis described in Chapter 4. Perhaps we are reading too much into this and the true explanation is altogether much simpler. The amounts of placental hormones produced is nothing more than a reflection of the robustness and health of the placenta, and by extension that of the embryo. Hormonal quantity then equates with embryonic quality. The more the better. They permit the mother to decide whether to terminate or continue with her pregnancy. This is the message that placental hormones are sending to the mother.

There is more. In addition to hormones, the placenta also secretes a variety of other products whose functions are presently unknown. They are labelled together as a family of 'pregnancy-specific glycoproteins (PSG)'.[152] Glycoproteins are a group of proteins that have a carbohydrate as a non-protein component. These PSGs are the most abundant placental-derived glycoproteins found in the blood of pregnant women, far more than the hormone HCG. They can be detected as early as 3 days post-fertilization, which is the time the blastocyst attaches to the uterine wall. Their concentration in maternal blood increases exponentially over the course of pregnancy, reaching the highest level at the third trimester. Low levels of PSGs are associated with pathological conditions such as miscarriage, intrauterine growth restriction, or pre-eclampsia. Experimental disruption of PSGs causes abortion in laboratory mice and monkeys. PSGs seem to be produced exclusively by the haemochorial placentas of rodents and primates and not by other types of placentas. There are 11 members of the PSG gene family in humans and 17 in mice. The independent expansion of the rodent and primate PSG gene families from a common ancestor supports a convergent evolution, implying that PSG function has been conserved across the two divergent species. Collectively, these observations indicate a critical role for PSGs in the maintenance of pregnancy. And yet, we do not know what they do.

In cell biology, a usual starting point in determining the function of a molecule is to identify its receptor. A molecule becomes active by initially binding to a receptor on a cell, like opening a lock with a key, the key being the molecule and the receptor the lock. By the same analogy, only the right key will open a particular lock. Scientists describe this as specificity. The next step is to find which cells in the body have this receptor as this will tell us where the molecule is likely to act. The molecule and the cells expressing the receptor are then put together in the laboratory and the cellular products resulting from this interaction analysed. These products will give us some idea as to the potential functions of the original molecule.

Using the procedures just described, the receptor for PSG has been identified as a molecule designated as CD9, at least in mice. This receptor is expressed by several cell types belonging to the immune system, so they are likely to be the targets for PSG action. When cultured together in the laboratory, PSG induces these immune cells to secrete a trio of proteins, namely IL-6, IL-10, and $TGF_{\beta1}$, which are usually referred to as 'non-inflammatory cytokines', so called because they tend to damp down or moderate the immune response to prevent it from being too aggressive. The overall conclusion based on these experiments is that PSG might act as an immune moderator by inducing cells of the immune system, via binding to their CD9 receptors, to secrete non-inflammatory cytokines that will protect the placenta from rejection by the mother. In addition to this role, some of the cytokines are involved in the growth of new blood vessels. PSGs, then, could have a dual function as an immune moderator as well as influencing the development of blood vessels at the implantation site, which will aid implantation.

Occasionally, products are found in the placenta that are totally unexpected, raising the inevitable question: why is the placenta making them? One such product is the neurotransmitter serotonin, also known as the 'happiness hormone' (although it is not strictly a hormone) because it affects the mood of a person and gives a feeling of well-being.[153] As a neurotransmitter, serotonin is involved in the transmission of nerve impulses. Released by one nerve cell, serotonin passes on the message to another nerve cell and so on, so that a chain of information is generated. If there is not enough serotonin, nerve cells cannot 'fire' properly and the message is not able to get through. This is a major cause of depression. Normally, nerve cells recycle serotonin by soaking it up to be released again, a mechanism known as 're-uptake'. A drug commonly used to treat depression belongs to the group of Selective Serotonin Re-uptake Inhibitors (SSRIs) and acts to stop this re-uptake so more serotonin is available. Serotonin is synthesized by nerve cells in the brain which is where it is needed. Now, to everyone's surprise, it is found also to be made by syncytiotrophoblast of the placenta. In addition, high levels of monoamine oxidase A (MAO-A) are also produced by the syncytiotrophoblast. This is an enzyme that destroys serotonin. Another group of anti-depression drugs are the Monoamine Oxidase Inhibitors (MAOI) which act to prevent this enzyme from degrading serotonin.

Since serotonin and MAO are made by syncytiotrophoblast at a critical time of fetal brain development, the placenta will exert a powerful influence on this development. Will anti-depressants given to pregnant mothers have any effects? These drugs can cross the placenta. As we discussed in Chapter 5 on fetal programming, any disruption will linger into adulthood. Are the seeds of depressive illness already sown during intrauterine life?

The placenta's impact on the fetal brain continues even after birth.[154] The pinnacle of human evolution is the dramatic increase in brain size, especially of the forebrain or neocortex which is responsible for higher functions, such as speech, thought, sensory, and motor activities. Much of this expansion takes place after birth, at a time when the mother develops a close bond with her young. The pattern of nerve cell interconnections in the newborn's maturing neocortex will be shaped by the quality of care the mother provides in this postnatal environment, in the same way as fetal programming is affected in the womb. As we saw in Chapter 4, maternal nurturing behaviour is triggered by placental hormones acting on her hypothalamus. In this way, the placenta is closely involved in ensuring the proper construction of the baby's neocortex, a vital part of the brain that endows us humans with our distinctive cerebral qualities. The placenta makes us what we are.

The human placenta synthesizes an eclectic collection of products, some of which are well characterized with established functions while others remain rather mysterious. Like Aladdin's cave, it is full of untold riches waiting to be retrieved.

8

Gateway

Everything that is transferred between the mother and her baby during pregnancy must pass through the placenta. It sifts through all the information very carefully, allowing some to cross while preventing other information from doing so. The placenta does more than merely select. It actively encourages the transmission of substances from the mother it considers to be beneficial to the baby.[155] The usual description of the placenta as a 'barrier' does not do it justice as this conjures up the image of a passive, inert structure barring all that comes its way. The placenta is more sophisticated than that. A more appropriate term might be 'gateway'. The dominant component of the placenta which constitutes this gateway is the syncytiotrophoblast because it is the outermost layer in direct contact with maternal blood. It is also a syncytium forming a continuous layer with no intercellular boundaries, so everything that is transmitted between mother and fetus must go through the syncytiotrophoblast. There is no way to bypass it, unless it is torn or injured. The syncytiotrophoblast, therefore, exerts a powerful influence on what is transmitted across the placenta. It is the sentinel to the gateway.

The most common mechanism of transfer of substances across the placenta is by simple diffusion. It is used for the exchange of important gases such as oxygen and carbon dioxide. The major factor that affects the rate of transfer of these gases is the blood flow within the placenta so that any situation that slows this blood flow will seriously compromise gaseous exchange between the mother and fetus. For example, in the disease malaria, the parasites have a tendency to aggregate in the blood spaces within the placenta, blocking the circulation of blood and depriving the fetus of oxygen. This is one way by which maternal infection can harm the fetus without the disease actually being passed on. As pregnancy advances, there is progressive thinning of the layers separating the maternal from the fetal circulation due to several changes taking place in the structure of the chorionic villi.[156] First of all, the syncytiotrophoblast itself becomes thinner. Then the underlying cytotrophoblast layer largely disappears so that what were originally two layers of trophoblast are now reduced to one. As a result, the fetal blood vessels are now brought much closer to maternal blood. During the first few months of pregnancy, the thickness of the layers separating fetal and maternal blood is estimated to be $\simeq$ 0.025mm and this is reduced to less than 0.002mm in the later stages so it is ten times thinner. These changes will facilitate gaseous interchange as the fetus grows.

The human fetus is very efficient in conserving oxygen once the gas has come through the placenta from the mother. Fetal blood has a higher concentration of haemoglobin (Hb) than adult blood. Haemoglobin is the protein in red blood cells that carries oxygen. Furthermore, fetal Hb (HbF) has a different structure to adult Hb (HbA) which gives the former a greater affinity for oxygen, that is, it binds more readily to oxygen. Both these features augment the total amount of oxygen in fetal blood.

Carbon dioxide production by the fetus occurs at about the same rate as oxygen is consumed. This carbon dioxide is transferred back from the fetus to the mother by diffusion across the placenta in the same way as for oxygen in the reverse direction. Excessive carbon dioxide in blood reduces its oxygen carrying capacity. Smoking is a major culprit which exacerbates this condition. Pregnant women who smoke tend to have small babies due to oxygen deprivation.

Besides gases, many other soluble substances also cross the placenta by simple diffusion. Most drugs given to the mother will reach the fetus in this way except for those with large molecular weights like heparin or insulin. This is worth remembering in view of the disastrous teratogenic (teratogen = monster maker) effects resulting from the administration of thalidomide in the early 1960s whose consequences are still being felt today. Disabilities caused by the drug ranged from limb deformities to brain damage, blindness, and deafness. Many did not survive but those who did are now finding it more difficult to cope with life as they reach middle-age. This episode has made doctors extremely nervous about prescribing drugs to treat the nausea and vomiting of pregnancy, popularly known as 'morning sickness', which was what thalidomide was used for.[157] Although seldom life-threatening, this condition, nevertheless, can have a profound effect on the quality of women's lives. Sometimes the vomiting is so severe as to require hospital admission for rehydration and correction of electrolyte imbalance, a clinical condition termed 'hyperemesis gravidarum'. Anti-nausea drugs are constantly being produced, evaluated, and subsequently withdrawn due to fear of liability suits for fetal defects. Pregnant women frequently do need other drugs for a variety of reasons. They may suffer from medical conditions that require continual treatment even while they are pregnant, such as asthma, epilepsy, high blood pressure, or diabetes. In these cases, it is essential to bear in mind the potential toxicity to the fetus while treating the mother. This is particularly important if a 'cocktail' of several drugs is used, as in the anti-retroviral treatment of HIV-infected pregnant women, where protection of the unborn child must be weighed against potential harmful effects. Another drug given to the mother that can harm the fetus is the synthetic hormone diethyl stilboestrol (DES). Here the effects on the fetus only become apparent in adult life many years later after the initial exposure. The sorry tale of DES was described in Chapter 7.

Improving on simple diffusion, some substances are helped across the placenta by binding to a carrier protein on the surface of the syncytiotrophoblast, riding 'piggy-back' as it were on this carrier protein which, unsurprisingly, is called a 'transporter'. This is known as facilitated transfer and permits transfer across the placenta at rates far greater than can be achieved by simple diffusion alone. The transmission of amino acids, the building blocks for proteins, takes place in this way.[158] There are 20 active amino acid transporters on the surface of syncytiotrophoblast. Each one can transmit several amino acids of similar structures and each amino acid can use different transporters, so there is a great deal of overlapping activity. Hormones and growth factors produced by the placenta can influence the activity of the transporter system.

The transport of glucose from mother to fetus utilizes a transporter known as GLUT-1 and is dependent on the concentration gradient between maternal and fetal blood.[159] When the mother has diabetes and her blood sugar level is raised, this excess sugar is transferred across the placenta to the fetus, stimulating overgrowth of the fetus. This is why diabetic mothers tend to have unusually large babies or 'macrosomia', which is defined as any birth weight exceeding 4000 grams. Curiously, macrosomia still occurs even when the mother's diabetes is well controlled and her blood sugar levels are maintained within normal limits. This means that something else must be happening to contribute to the increased transfer of glucose to the fetus. Examination of syncytiotrophoblast from pregnancies complicated by maternal diabetes shows that the level of GLUT-1 transporters is increased compared to placentas from normal pregnancies. This leads to more glucose passing over to the fetus.

The ability to regulate the activity of transporters allows the placenta to act as a nutrient 'sensor' to coordinate transporter function with availability of nutrients.[160] When maternal blood is rich in sugar as in diabetes, the placenta senses that food is plentiful and will upregulate the GLUT-1 transporters to transmit more glucose to the fetus. On the other hand, when food is scarce transporters are downregulated, restricting the amount of nutrients that can be transmitted to the fetus. Fetal growth will slow as a result. In this way, the placenta matches what the fetus needs with what the mother can provide. It is better to have a small baby or even to lose a baby when food is not available and preserve the mother's life for her to breed again. Nature tends to favour the mother over the baby when there is conflict of interest. This makes sense in evolution.

Not all transporters function to transfer nutrients from mother to baby. Some actively *expel* substances from the placenta and pump them back to the maternal circulation instead of allowing them to cross. For the placenta to possess this type of transporter is unexpected, but they obviously play a very important role in protecting the fetus from noxious agents coming from the mother. One such transporter is the 'breast cancer resistant protein' (BCRP), so called because it was discovered in 1998 that it confers drug resistance by preventing entry of the drug into breast cancer cells. The placenta has ingeniously taken over this transporter system for its own use to prevent harmful drugs from reaching the fetus.

Substances the fetus needs from the mother that are too large to be transferred across the placenta by simple diffusion or by transporters can be taken into the syncytiotrophoblast by receptors on the cell surface. This is known as 'receptor-mediated transfer'. This mechanism requires energy. The prime example is the transmission of antibodies, also known as 'immunoglobulins' (Ig) from mother to fetus during pregnancy.[161] We shall take a closer look at this because the passage of the 'wrong' kind of antibodies can have dire consequences. Antibodies are special blood proteins, produced in response to the presence of an 'antigen' (usually foreign), to attack the antigen and render it harmless. Antigens are a diverse group, ranging from bacteria, viruses, pollen, cells, and so on. Hence, antibodies are important components of the host immune defence. This transfer of maternal antibodies protects the fetus against infections before birth as well as for the first few months after birth until the newborn's own immune system becomes fully competent.[162] This is especially important in areas of the world, such as the tropics, with a high prevalence of endemic diseases. Without the transfer of antibodies from the mother, many babies would not survive.

I shall now briefly describe the structure of the antibody molecule because this helps to understand how it is transported. The antibody molecule is shaped like the letter 'Y' (see Figure 15). The stem of the Y is known as the 'Fc' region. The short arms of the Y, known as 'Fab', carry the antigen recognition sites, that is, areas which will bind to the antigen to which the antibody is made. Subtle differences in the Fc region separate antibodies into five classes defined as: IgG,

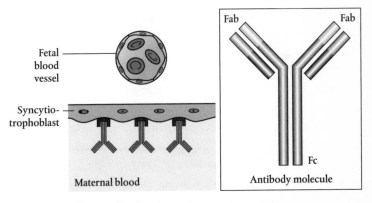

FIGURE 15 A Y-shaped antibody molecule (inset) binding to FcRn receptors on the surface of syncytiotrophoblast via its Fc region prior to transmission across to the fetal circulation.

IgM, IgA, IgD, and IgE. Of these classes, only IgG is transmitted across the placenta. The other classes of antibodies do not cross. This is because the syncytiotrophoblast layer of the placenta possesses special receptors which will bind only to IgG and not to any of the other classes. Since the Fc region is where the class of antibody is determined, binding to this region is how the receptor selects IgG for transmission. The receptor must be remarkably sensitive to subtle changes in the antibody Fc region for it to be able to distinguish between the different classes of antibodies. The receptor is called FcRn (FcR = receptor for Fc region; n = neonatal).[163] After binding to FcRn, the IgG molecule is protected from degradation inside the syncytiotrophoblast and is safely transported to the other side into the fetal circulation. The other classes of antibodies that are not bound to FcRn are all broken down inside the syncytiotrophoblast. The syncytiotrophoblast is specially designed for this task. On the surface of the syncytiotrophoblast, there are large numbers of small projections known as 'microvilli' which greatly expand the total area for absorption of antibodies. The transfer of IgG across the placenta is quite slow before the 24th week of pregnancy but, after that, the rate of transfer increases exponentially during the second half of pregnancy. At term, IgG levels in the fetus ($\simeq$ 15g/dl) even exceed those in the mother ($\simeq$ 13g/dl).

FcRn is an unusual and highly versatile receptor whose biological importance is beginning to be fully appreciated.[164] In addition to the placenta, FcRn is also located in several other organs in adults, such as the brain, kidney, and intestine. It is possible that the receptor might be involved in IgG transport in these organs in the same way as it is across the placenta. FcRn is also found on endothelial cells lining blood vessels. Because of its ability to bind and protect IgG from being broken down, the presence of FcRn in blood vessels could help to extend the life of IgG within the circulation. When IgG levels in the blood are low, more FcRn in blood vessels is available to bind and protect IgG so that more IgG is salvaged. Conversely, when

IgG levels are high FcRn becomes saturated, resulting in more IgG being destroyed because it is not bound and protected. In this way, FcRn functions not only as a receptor for IgG transport but also as a homeostatic control for the survival of this very important class of antibody. It is both a salvage and a transport receptor. Structurally, apart from a few small differences, FcRn bears a remarkable resemblance to the HLA molecule which we already encountered in Chapter 6 on immunology and graft rejection. Now we see a structurally similar molecule, FcRn, performing a totally different function for IgG transport and survival. This is an interesting illustration of how one ancestral molecule can evolve into two with quite distinct functions without too drastic a change in its basic structure. Nature is economical in this way.

Efficient as it is, this antibody transport system is not foolproof. While it can select IgG from the other classes of antibodies, it cannot distinguish between 'good' IgG and 'bad' IgG. This is due to the way the antibody molecule is transmitted. As we have already described, the antibody crosses the placenta 'backwards' as it were by using its Fc region, which identifies it as IgG, to bind to the FcRn receptor on syncytiotrophoblast. The short arms of the Fab portion of the antibody molecule, which determine its reactivity, have no influence on the transmission. That is the system's Achilles heel. Good IgG are those antibodies which will protect the baby against infections while bad IgG are those made by the mother against incompatible components in the baby derived from the father. The latter, when it crosses the placenta, can harm the baby. The best known example is haemolytic disease of the fetus and newborn due to incompatibility between mother and baby for the Rhesus blood group.[165]

The Rhesus blood group is an antigen found on the surface of red blood cells, like the ABO blood groups. It is so called because of its initial discovery in the Rhesus monkey. It is really a group of antigens Rh-C, -D, and -E of which the most important is Rh-D because this antigen is already well developed in the red blood cells of the fetus and the mother is easily sensitized to it and will produce antibodies against it. Hence, discussions on Rhesus incompatibility usually centre on Rh-D. The human population is divided into those who have this antigen on their red blood cells (Rh-D⁺) and those who do not (Rh-D⁻). If an Rh-D⁻ mother bears an Rh-D⁺ fetus who has inherited the antigen from an Rh-D⁺ father, she will make antibodies against the fetus' Rh-D antigen because it is foreign to her. The stimuli for this antibody production are red blood cells from the fetus that have leaked across into the mother's circulation. Unfortunately, the Rh-D antibodies made by the mother belong to the IgG class which will, therefore, readily cross the placenta. When these antibodies reach the fetal circulation, they will destroy the fetal red blood cells because these antibodies are designed to attack the foreign Rh-D antigen. The result is haemolytic disease of the fetus or newborn depending on how early the haemolysis occurs (haem = blood, lysis = destruction). The placenta is usually an efficient barrier against the transmission of red blood cells during pregnancy so the bulk of these sensitizing fetal cells would have crossed at birth when the placenta separates from the uterus. This is why an Rh-D⁻ mother tends to produce Rh-D antibodies only with the second or subsequent babies and not the first.

Readers will be familiar with the more common ABO blood groups and the disastrous results that follow from mismatched transfusions when the donor is not of the same blood group as the recipient. What happens when a mother's ABO blood group differs from the baby's? Does this lead to haemolytic disease of the fetus in the same way as occurs in Rh-D incompatibility? Fortunately, the answer is 'no'. This is because ABO antibodies in blood are of the IgM class and, therefore, cannot readily cross the placenta to affect the fetus. ABO antibodies are known as 'natural' antibodies because everyone has them. We are not born with these antibodies but acquire them very early on after birth. The human newborn enters the world sterile but is rapidly colonized by bacteria in the gut during the first few months of life. These bacteria are thought to stimulate the neonate to produce ABO antibodies.

There are, however, some mothers whose ABO antibodies are IgG instead of IgM but even these IgG antibodies are not as damaging to the baby as Rh-D antibodies. There are many reasons for this. Although the ABO antigens are, by common usage, known as 'blood groups' their distribution is, in fact, far more widespread than the name would suggest. ABO antigens are ubiquitous and are found in most fetal and placental cells and not just on red blood cells. So, even when IgG ABO antibodies succeed in crossing the syncytiotrophoblast, they are 'mopped up' by other cells within the placenta that have ABO antigens on their surface. The placenta can act as a kind of 'immunological sponge', absorbing out potentially harmful ABO antibodies before they can reach the fetus.[166] Even if some of these antibodies succeed in reaching the fetus, they are further absorbed out by other fetal cells. Unlike ABO antigens, the Rh-D antigen is found only on red blood cells so that when Rh-D antibodies reach the fetus, the fetal red blood cells will bear the full brunt of the attack by these antibodies. Furthermore, ABO antigens are not fully developed in the fetus, unlike Rh-D which is present very early, so there are fewer antigenic targets to be attacked by ABO antibodies. All these features explain why full blown haemolytic disease is relatively uncommon even if there is ABO incompatibility between mother and fetus. If it occurs at all, the disease tends to be mild, presenting as slight jaundice which we see in some babies due to destruction of a few fetal red blood cells.

Fetal–maternal ABO incompatibility could even have a protective effect against haemolytic disease resulting from Rh-D incompatibility. Any Rh-D⁺ fetal red blood cells that have gained access to the mother's circulation would be rapidly destroyed by her ABO antibodies before they have time to stimulate her immune system to make antibodies against Rh-D. There would be fewer cases of haemolytic disease caused by Rh-D incompatibility if fetus and mother are also incompatible for ABO blood groups compared to those who are ABO compatible. This is indeed what happens. In other words, in matings which are both Rh-D and ABO incompatible, there is a significantly reduced risk of babies affected by haemolytic disease. This is an interesting example where incompatibility for one antigen (ABO) can protect against the harmful effects due to incompatibility of another antigen (Rh-D).

Based on the above observations, a technique has been devised which consists of artificially administering Rh-D antibodies at 28–34 weeks gestation routinely to all pregnant Rh-D⁻ women who are married to Rh-D⁺ men, the rationale being to destroy all Rh-D⁺ fetal red blood cells that may have gained access to her circulation before they have the chance to stimulate her to produce Rh-D antibodies.[167] This regime is now the standard procedure and has dramatically reduced the incidence of haemolytic disease of the fetus and newborn. This is one of the great success stories resulting from our understanding of antibody transport across the placenta. But haemolytic disease has not been completely eradicated by this treatment. In England and Wales, about 500 fetuses continue to develop this disease annually. Also, it is estimated that fetal loss before 28 weeks of pregnancy due to haemolytic disease is responsible for a significant number of spontaneous abortions each year. Why is this? The reason is because there are other red blood cell antigens besides Rh-D that can be troublesome. The International Society of Blood Transfusion recognizes over 302 red blood cell antigens. Many of these are potentially capable of causing haemolytic disease because they can stimulate the pregnant mother to make IgG antibodies which then cross the placenta to destroy fetal red cells in the same way as Rh-D antibodies. Because anti-D prophylaxis has been so successful, these so-called 'non-D' antibodies are coming out of the woodwork and becoming more important as a cause of haemolytic disease. The lesser known Kell blood group is one of these.

The Kell blood group may soon emerge from under the shadow of Rhesus to claim a rightful place in history. It may even be featured in a television Tudor period drama in the future! Medical historians have long suspected that the sequence of reproductive failure by Henry VIII might have been caused by blood group incompatibility between him and his multiple wives. Henry initiated at least 11 and possibly 13 pregnancies so he was not infertile. The problems came later with frequent fetal and neonatal mortality. This pattern of late pregnancy loss fits in well with that seen in Rhesus incompatibility. The historians Whitley and Kramer have argued that the source of Henry's woes is the Kell blood group and not the Rhesus blood group.[168] Both these blood groups cause haemolytic disease of the fetus and newborn in a similar way, so that what we have already described for Rhesus is equally applicable to Kell. Henry would probably be Kell positive while his wives were Kell negative. The distribution of the Kell blood group is such that the vast majority of people are Kell negative and only a few are Kell positive, so Henry would have had a good chance of meeting up with a Kell negative wife. The rare Kell positive woman in the population is not compromised since she will not react to either a Kell positive or Kell negative husband. Neither will trigger antibody production in her that might harm the baby. It is the rare Kell positive man who causes incompatibility problems when mated to a Kell negative wife. He is the rogue male. Henry could be such an example. The argument for the villain of the piece being the Kell blood group is the observation that Henry also suffered from a variety of psychological symptoms and personality changes in his later years that are characteristic of McLeod syndrome, a medical condition to which some Kell positive patients are susceptible.

The story of Henry's quest for an heir began with the tragic obstetric history of his first wife, Catherine of Aragon. If Rhesus or Kell prevention were available to Catherine in those days, then the whole of English history might have been different. The saga of Henry's multiple wives would not have occurred if Catherine had borne him a son. There was just one surviving offspring to their union who was a girl (Mary), much to the disappointment of Henry who so much wanted a male heir to the throne. This raises an interesting question. Are the effects of Rhesus or other blood group incompatibility more pronounced in male fetuses compared to female fetuses? A male has the 'Y' chromosome which the female does not possess. In terms of transplantation immunology, this 'Y' antigen could produce a 'non-self' signal to the mother which will not occur with a female fetus. The combined 'foreign' signals generated by the 'Y' and blood group antigens together would be stronger than that arising from each antigen alone. There are reports that among the children born to Rhesus negative mothers who developed haemolytic disease there are more boys than girls, confirming that there seems to be a stronger maternal immune response when the fetus is incompatible with the mother for both the 'Y' and Rhesus antigens.

There has been a great deal of debate over the years as to what effects the 'Y' antigen has on human reproduction. Does it make any difference to reproductive success? The human sex ratio (M:F) at birth is approximately 109:100, that is there are slightly more boys born than girls.[169] But if this figure is projected backwards to earlier in gestation using material from elected terminations, then the ratio can be as high as 166:100. What this means is that there is a greater differential loss of male fetuses *in utero* between conception and birth. The Y-bearing male sperm is thought to swim faster than the X-bearing female sperm in the race up the cervical canal, through the uterine cavity, and into the Fallopian tube where it meets up with an egg coming from the opposite direction after having been shed from the ovary. This could explain the preponderance of boys conceived. But from that point onwards it is all downhill for the male fetus. The female fetus seems better able to survive in the maternal uterus than the male. What might be the explanation?[170] Since the environment inside the uterus is the same for the fetus of either sex, this differential viability could be due to the 'non-self' Y antigen of the male giving out foreign signals to the mother. There is considerable evidence from studies in inbred mice that the antigens controlled by the Y chromosome do have a significant impact on transplantation. For example, female to male skin grafts are usually accepted but similar grafts from male to female are rejected. Inbred mice are engineered to be genetically similar to each other so that any slight difference, like the presence of an extra Y antigen in a male, will show up. But in an outbred human population, there are so many disparate antigens among individuals that any effects produced by the Y antigen will be masked by the overall background genetic 'noise' and, thus, difficult to detect.

In China today, the M:F birth ratio is as high as 124:100, presumably as a result of the one-child policy leading to the preferential abortion of female fetuses. It is estimated there could be almost 100 million Chinese males with no females to mate with. How this artificially skewed birth ratio will affect Chinese society in the future remains to be seen. A similar situation has developed in India where the availability of cheap ultrasound scans to determine prenatal sex has led to the abortion of female fetuses on an unprecedented scale. Half a million girls are said to be aborted every year in India, more than the total number born annually in Britain. This has altered the natural sex ratio at birth dramatically.

The frequency of the Rh-D antigen varies in different populations. In Caucasians 12–18 per cent of individuals are Rh-D⁻ while the figure for Arabs is 10 per cent and for Africans 2–5 per cent. Interestingly, people from East Asian countries such as China and Japan are almost all Rh-D⁺ with hardly any Rh-D⁻ individuals. Because of their relatively high percentage of Rh-D⁻ individuals, haemolytic disease of the fetus and newborn is encountered mostly in Caucasian populations. Biologists are always intrigued as to why certain genes are conserved within a population while others are discarded during evolution. It is logical to think that this is because some genes are useful for survival while others are not. A good example among blood groups is the Duffy blood group. Duffy acts as a receptor for a certain species of malaria parasite, allowing the parasite to enter the red blood cell which is its usual habitat. Without Duffy, the parasite cannot enter the red blood cell. Individuals lacking the Duffy blood group will, therefore, be resistant to infection by this species of malaria so that, as a result of natural selection during evolution, Duffy becomes less common in ethnic groups from areas of the world with high levels of this kind of malaria. Many individuals in Africa are indeed Duffy⁻. We do not know what the selective advantage might be of being Rh-D⁻ nor why the proportion of individuals who lack this antigen should vary in different populations since being Rh-D⁻ is so problematic in reproduction.

Red blood cells are not the only cells that can be injured by maternal antibodies. Two other components in fetal blood can also become targets for damage. These are the white blood cells (neutrophils) and the platelets. Like red blood cells, fetal neutrophils and platelets also carry surface antigens derived from the father which are not present in the mother so the latter will make antibodies against these antigens in the same way as she does against fetal red blood cells. These antibodies also belong to the IgG class and will readily cross the placenta.

Neutrophils form our first line of defence against infective organisms, particularly bacteria. They roam the body in search of these invaders and, when found, will engulf and ingest them. This is why neutrophils are also known as phagocytes: cells that eat. Phagocytosis is a very ancient trait. Primitive unicellular organisms like amoebae ingest their food in this way since they have no gut. Since then, this propensity has been conserved in higher vertebrates by certain cells like neutrophils and used to defend the body against invading pathogens. Phagocytosis has been transformed from a nutritive to a defence function. Neutrophils possess powerful intracellular enzymes that can digest bacteria once they have been ingested. Pus, which forms over infected areas, is made up of a collection of neutrophils. In recognition of its protective function, 'pus, pus, laudable pus' is a familiar battle cry among pathologists. If a baby's neutrophils are damaged or their numbers reduced by maternal antibodies, they will be susceptible to infections. These tend to be local infections, mainly of the skin or mucous membranes, but life-threatening infections, such as pneumonia or meningitis, can sometimes develop resulting in significant mortality. Fortunately, this period of increased susceptibility to infections is transient because the maternal antibodies are 'foreign' proteins and will be progressively degraded within a few weeks and the baby's neutrophils will soon recover.

Platelets are involved with blood coagulation. A decrease in platelets in the blood is known as thrombocytopenia. Without adequate platelets, blood will not clot properly. Damage to fetal platelets by maternal antibodies will then lead to symptoms of clotting defect in the baby. These may be mild, such as pin-point spots of bleeding under the skin (petichiael haemorrhage) or larger areas of bruising (purpura) but severe bleeding internally can also occur. This can happen *in utero* or after birth and is responsible for significant morbidity and mortality in the affected infants. In particular, great care must be taken during delivery to avoid trauma to the head as this could lead to fatal intracranial haemorrhage. These conditions affecting platelets are more common than those for neutrophils and are seen in approximately 1 in 1000–2000 live births.

All the above examples are diseases caused by the transmission of IgG 'allo-antibodies' across the placenta. The prefix 'allo' denotes antibodies which are directed at antigens present in another individual. Hence, maternal antibodies produced against fetal red blood cell, neutrophil, or platelet antigen inherited from the father which the mother does not possess are 'allo-antibodies'. But there is another group of IgG antibodies known as 'auto-antibodies'. The prefix 'auto' defines antibodies which are directed at 'self' antigen. Many diseases are caused by an individual producing an inappropriate immune response against 'self'. These diseases are called autoimmune diseases. (See Chapter 6.)[171]

Thrombocytopenia is one such disease. A mother with autoimmune thrombocytopenia will possess auto-antibodies against her own as well as her baby's platelets, whereas if she has developed allo-antibodies these will react only against the baby's platelets. Therefore, if the condition is caused by auto-antibodies, both the mother and baby will present with thrombocytopenia whereas allo-antibodies will affect only the baby.[172] Another example is Grave's Disease or hyperthyroidism which is due to the production of auto-antibodies by the patient that overstimulate her own thyroid. A mother suffering from Grave's Disease will have these auto-antibodies. Being IgG, they will be transmitted to the fetus across the placenta. Furthermore, the antigens in Grave's Disease that are the targets for the auto-antibodies are present only in the cells of the thyroid gland. They are referred to as 'organ-specific', that is present only in that particular organ. The 'mopping up' process by the placenta described earlier cannot take place because there are no equivalent thyroid cells in the placenta to fulfil this role. When transmitted to the fetus, the fetal thyroid will bear the full brunt of the effects of this auto-antibody and the baby will be born with signs of hyperthyroidism like those in the mother. At one time, this association between mother and child was mistakenly attributed to genetic inheritance of the disease passed down from one generation to the next. This illustrates how mechanisms of disease production can easily be confused.

One more unwanted side effect of antibody transfer is that some antibodies produced by the mother against infections might carry these infective organisms bound to the antigen-recognition sites on their Fab arms. When these antibodies are transported across the placenta via the receptor FcRn, these infective organisms may also be inadvertently carried over to the fetus. This has been described as the 'Trojan Horse' mechanism, as in the story of Helen of Troy where Greek soldiers sneaked into Troy hidden inside a large wooden horse. Some drugs might also use this mechanism for transmission. For example, insulin is normally too large a protein molecule to cross the placenta. Being a foreign protein, a diabetic mother treated with insulin occasionally makes IgG antibodies against this drug. Bound to its antibody and hidden inside its Trojan Horse, insulin can unsuspectingly find its way from mother to baby.

From the preceding survey, it is clear that the receptor-mediated transfer of IgG across the placenta is not perfect by any means. This is because the IgG bound to the FcRn receptor is transmitted via its Fc region so there is no way for the placenta to discriminate between beneficial IgG for protection against infections and harmful IgG directed at fetal components, such as red blood cells, neutrophils, platelets, and the thyroid, or those antibodies harbouring stowaway infective organisms. All these antibodies just have to be lumped together as IgG and allowed to cross. Nevertheless, the benefits of conferring immunity to the fetus *in utero* far outweigh the occasional unwanted side effects. Nature does not always have clear-cut solutions to every problem.

While on the subject of antibody transfer from mother to young, it should be mentioned that maternal antibodies can also reach the baby after birth via milk.[173] In contrast to IgG antibodies, which are transferred across the placenta, those in milk are mostly IgA. This class of antibody has a special structure which allows it to be secreted into milk by the breast. Interestingly, these milk antibodies are mostly directed at bacteria or viruses which have previously infected the gut of the mother. These antibodies then line the gut of the infant when it ingests its mother's milk, protecting it from any gastro-intestinal infective organisms it may encounter during its first few months of extra-uterine life. We can appreciate how the process is beautifully adapted for the intestinal immune experience of the mother to be used to protect the gut of the next generation via milk. In many animals, suckling after birth is the only means of getting antibodies from the mother. There is no antibody transfer across the placenta particularly in those animals with epitheliochorial placentas. Presumably, there are too many layers between fetal and maternal circulations to permit efficient antibody transfer. This is a major disadvantage in having an epitheliochorial type of placenta.

The human offspring, then, is extremely fortunate in having protection by maternal antibodies before birth via transfer of IgG across the placenta and also after birth by IgA in milk. Not to breastfeed babies might be depriving them of a valuable means of defence. To breastfeed or not to breastfeed is an issue which continues to be hotly debated. It cannot be denied that breast milk does contain some rather useful immunological protective ingredients. According to the United Nations Children Fund (UNICEF), formula-fed infants are more susceptible to gastroenteritis than those who are breastfed. Breastfeeding could save over 700 infant deaths per year in the United States. After all, we are mammals and the predominant trait in mammals is to lactate and suckle their young. Why do we want to overturn 250 million years of evolution? Even our primitive cousin the duck-billed platypus does it. Breastfeeding has an additional advantage. Suckling an infant delays the onset of the next pregnancy and thereby increases the birth intervals. This is Nature's own method of contraception. There has been a proposal to expand the human milk donation service to the same level as for blood donation. This will allow women who cannot breastfeed their infants to have access to human milk.

The growing fetus needs large amounts of minerals, such as iron and calcium. The human fetus needs up to 5mg of iron per day, all of which is derived from the mother. Iron is normally bound to a protein, transferrin, circulating in the mother's blood. This iron coupled to transferrin in the maternal circulation is transferred to the fetal circulation across the placenta after binding to a transferrin receptor on the surface of syncytiotrophoblast. This is analogous to the transfer of antibodies by FcRn. Once the complex has crossed the placenta, the iron molecule is released from the transferrin carrier. There is a parasite called schistosome which belongs to the family of flukes. It causes the very debilitating disease schistosomiasis in the tropics. The parasite has an ingenious way of evading the host immune response by covering itself with proteins absorbed from the host's own blood, thereby disguising itself as 'self'. So successful is this camouflage that the worm can live happily inside the host for years and even decades. This observation had led some reproductive immunologists to propose that absorption of maternal transferrin onto the surface of syncytiotrophoblast might serve a similar purpose.

Calcium is required for fetal bone development. Unlike iron, calcium is transferred from mother to fetus across the placenta using calcium transporters, such as PMCA3, in the same way as for aminoacids and glucose. Lipids are also transferred in this way using a transporter CD36.

The placental form of mammalian reproduction is highly successful because the fetus can be nourished safely by the mother for long periods in the relatively germ-free internal environment of the uterus. In order for the placenta to protect the fetus, it should be able to prevent the transmission of infective organisms from the mother. How successful is the placenta in fulfilling this task? The fact that intrauterine infections, such as rubella, with its teratogenic effects on the baby, can occur means that the placenta is not totally impermeable. But we must not rush too hastily to blame the placenta because infective organisms from the mother can reach the baby by other routes besides going through the placenta. They can affect the baby directly from infection in the abdomen or the pathogens can arrive via the vagina and cervical canal in what is termed 'ascending infection'. Many bacterial intrauterine infections use this route of transmission. The placenta cannot be expected to protect the baby against such infections.

Most maternal infections do not pass to the fetus, so the placenta does a good job in preventing the passage of pathogenic organisms. Visual evidence of this efficient placental barrier can be seen in malaria. In this disease, huge numbers of parasites amass in the blood spaces of the placenta. This was observed by early investigators from the beginning of the 20th century and referred to as 'placental malaria'. The sluggish blood flow inside the placenta, together with the relatively low oxygen concentration in placental blood, favour the survival and replication of the parasites. Also, the red blood cells infected with parasites are sticky and attach themselves to a molecule called chondroitin sulphate A on the surface of syncytiotrophoblast, so the parasites are effectively trapped in this location. Yet, in spite of this intense concentration of parasites within the placental blood spaces, there is little damage to the placenta itself and, more importantly, the parasites do not manage to breach the placental barrier and reach the fetus. Malaria demonstrates another protective mechanism which the fetus has up its sleeve. Malarial parasites need to live inside red blood cells containing adult haemoglobin (HbA). They cannot survive inside fetal red blood cells containing fetal haemoglobin (HbF) which has a different structure. Hence, any transmitted parasites will find the fetal environment less hospitable than the one they are accustomed to in the mother and will not survive for very long. The inability of the parasites to cross the placenta combined with their failure to thrive in the fetus explains why congenital malaria is extremely rare.

In spite of being protected from the malarial parasites, the fetus does suffer. The mass of parasites in the blood spaces interferes with placental blood flow and, hence, the transfer of oxygen and nutrients from mother to fetus. Because of this, patients with malaria tend to give birth to small babies. Severe cases can even result in miscarriage and still birth. One of the characteristic symptoms of malaria is anaemia because the infected red blood cells are damaged by the parasite. This maternal anaemia will also lead to poor fetal nutrition. In addition, red blood cells can be destroyed by the patient's own immune response producing antibodies against the infected red cells in an attempt to get rid of the parasites within them. These antibodies cannot discriminate between infected and uninfected red blood cells, so will destroy both. This explains why the degree of anaemia in a patient with malaria can be very severe even when the patient is only lightly infected with the parasites. The patient's own antibodies are responsible for the red blood cell destruction, not the parasites. Unfortunately, these anti-red blood cell antibodies are IgG which are freely transmissable across the placenta and will destroy fetal red blood cells once they reach the fetal circulation. The fetus can, therefore, suffer from anaemia without actually being infected with malaria. This is another instance when the active receptor-mediated transport of antibodies across the placenta is insufficiently discriminatory to select out the good from the bad. The lesson to be learnt here is that maternal infections can affect the fetus indirectly without actually producing overt infections in the latter.

The syncytiotrophoblast is undoubtedly the key player in protecting the baby from maternal infections. Being a syncytium with no intercellular boundaries, it forms a continuous frontier between the fetal and maternal circulation. It is, itself, relatively resistant to infections. If, for any reason, the integrity of the syncytiotrophoblast layer is breached by infective organisms, the placenta can call on other cells to deal with them. These are phagocytes that can engulf and digest infective organisms. In the placenta, they are known as 'Hofbauer' cells, named after the German scientist who first described them.[174] These cells will come into play as a second line of defence. Certain other organs of the body such as the liver, spleen, and lymph glands are also populated by similar phagocytic cells. These organs are collectively part of what is known as the 'mononuclear–phagocyte' system whose job it is to survey the internal milieu of the body for invading organisms. The placenta also belongs to this system, and its role is to defend the fetus from infective organisms coming from the mother. We noted in an earlier chapter that the placenta is part of the endocrine system because of its hormone production. Now, we see that it is also a member of the mononuclear–phagocyte system by virtue of its defence function. This illustrates the remarkable versatility of the organ.

Although the combination of the syncytiotrophoblast layer and the presence of phagocyte cells within the placenta constitutes a formidable barrier, a few organisms have cleverly devised ways to circumvent this defence. If we make a list of the infective organisms that most frequently cause intrauterine infections, they are many viruses (rubella, cytomegalovirus), a bacterium (listeria), and two protozoan parasites (toxoplasma and South American trypanosome). What these pathogens have in common is they are all intracellular organisms, meaning they live inside the cells of the host they have infected. Take, for example, the protozoan parasite *Toxoplasma gondii*, the cause of the disease toxoplasmosis. Toxoplasma have evolved a mechanism to resist intracellular digestion after they have been ingested by placental phagocytes. They can even grow and multiply within these cells. Thus, not only are they not killed, they have now created for themselves a nice micro-environment which permits their replication and from where they are sheltered from the host's other defence armoury such as antibodies. It is this ability that allows toxoplasma from the mother to cross the placental barrier to infect the fetus. This protozoan, then, is an important cause of intrauterine infection. The effects on the fetus will depend on the gestational age when infection occurs and the number of organisms which succeed in reaching the fetal circulation. These effects range from a mild illness to malfunctions of the eye and the central nervous system which are characteristic of congenital toxoplasmosis. Even intrauterine death can occur.

T.gondii is a typical zoonosis, meaning a disease which is freely transmissible between animals and humans with a lifecycle that includes humans and other animal species such as cattle and cats. Meat from infected cattle and excrement from cats harbour living forms of *T.gondii*. The advice to pregnant women not to eat poorly cooked beef and to avoid close contact with cats has a sound scientific basis. France has a higher rate of infection ($\simeq$ 10 per 1000 pregnancies) compared to the UK ($\simeq$ 2.3 per 1000 pregnancies) which may be due to the inclusion of rare meat, such as steak tartare, in French cuisine. Not all cases of maternal toxoplasmosis necessarily give rise to fetal infection. It is estimated that about 1 in 3 cases do so. Something, therefore, is protecting the fetus. The placenta is likely to be responsible for this protection in spite of the parasite's evasion strategy.

A further example that may be quoted as illustrating the importance of placental phagocytic cells is the disease trypanosomiasis, caused by two distinct species of the protozoan parasites. African trypanosomiasis is not normally associated with intrauterine infection, whereas the South American disease shows evidence of transmission from mother to fetus. The cardinal difference in the lifecycle of these two protozoa is that the South American trypanosomes have adopted an intracellular phase in their development which is not present in the African variety. The former has evolved a mechanism for survival inside a phagocytic cell whereas the latter has not learnt to do so.

We must not forget that the fetus itself is not entirely defenceless against pathogens that have managed to breach the placental barrier. The human fetus can generate its own immune response as early as 21 weeks of gestation, although this has not yet reached its full adult potential.[175] Children with congenital infections, such as toxoplasmosis, frequently have raised levels of IgM antibodies in their blood directed specifically at the toxoplasma organism.[176] Since IgM antibodies do not cross the placenta (only IgG antibodies do), these antibodies must be produced by the fetus and not transferred from the mother.

The ability of the fetus *in utero* to mount an active immune response can be usefully employed clinically. We could immunize a fetus while it is still in the womb by administering the vaccine to the mother. This would be particularly handy in those areas of the world where infectious diseases early in life are important causes of infant mortality. This idea was raised way back in the early 1970s but died out.[177] It has now been revived by the observation that infants born to mothers who receive the flu vaccine are less likely to come down with flu than those born to mothers who did not have the vaccine. This has led to the recommendation that all pregnant women should be given the flu vaccine, thereby protecting themselves as well as their offspring. Although the protection afforded to the baby is assumed to be by the transfer of maternal antibodies across the placenta, the possibility of the fetus generating its own active immune response to the vaccine should not be dismissed. Either way, the procedure will benefit the offspring.

If the above recommendation is adopted, it will set a new trend in neonatal care. Inevitably, there will be many questions. Should vaccination against other diseases be considered? What kind of vaccine is best: one that makes the mother produce antibodies which will be transferred across the placenta to protect the baby or one that will actively stimulate the baby's own immune system? Would a vaccine containing dead organisms be safer than one containing live ones? Timing is also important. At what period during pregnancy should the vaccine be administered to take full advantage of the gradual maturation of the fetal immune system? This will be a fruitful area of research collaboration between paediatricians, obstetricians, immunologists, and placental biologists.

The human placenta is relatively impervious to the passage of cells. If this does occur, it is usually at birth when the placenta separates from the uterus. This is why haemolytic disease of the fetus and newborn usually affects only the second or subsequent pregnancy, because enough fetal cells have crossed to sensitize the mother only during the birth of the first baby. However, a small number of fetal cells do occasionally manage to transverse the placenta during pregnancy, probably through small tears in the syncytiotrophoblast. In animal studies, more fetal cells are seen in syngeneic (fetus and mother genetically similar) compared to allogeneic (fetus and mother genetically different) pregnancies. The genetic and immunological relationship between fetus and mother is therefore an important determinant for the transplacental passage of cells and their ultimate survival. This is not surprising because fetal cells that are genetically alien to the mother will be more quickly destroyed by the mother's immune system and do not survive as readily as cells that are compatible. We do not know if this also applies to humans although there is no reason why it shouldn't.

There is a lot of interest in these transmitted fetal cells and a great deal of effort is being made to find them in maternal blood. There are two reasons why these cells are of interest. First of all, because they contain fetal genetic material they will be very useful for prenatal screening to determine whether a baby is at risk of genetic disorders, such as Down's syndrome. A 20 year old woman has about a 1 in 1500 chance of having a baby affected by this disease while the risk is raised to 1 in 100 in a 40 year old. The incidence of Down's syndrome has increased dramatically in the past few decades as more women opt to postpone having families, so it is important to have a reliable method of screening women of susceptible age. At present, there are two ways to obtain fetal cells for this purpose. One is by sucking out a bit of amniotic fluid in the womb which will contain some fetal cells (amniocentesis). The other method is by snipping off a small piece of placental tissue through the cervix (chorionic villous sampling). Unfortunately both these methods are not without danger and could disrupt the pregnancy. Such invasive techniques carry a 1 per cent risk of inducing miscarriage, which is the same level of risk as for a 40 year old mother having an affected baby. They should not be carried out without due consideration. For this reason, doctors are trying to find an alternative non-invasive test. If fetal cells can be isolated from maternal blood, then this could be the ideal answer. Getting blood from the mother by venepuncture is a routine procedure and should not interfere with pregnancy in any way. While the idea is a good one, it has proved extremely difficult in practice because of the very small number of cells present. Even the use of an HLA-G antibody to identify trophoblast cells has not been too successful (see Chapter 6).

Attempts to isolate fetal cells are now, more or less, superseded by more modern technology whereby fetal genetic material itself in the form of DNA or RNA can now be extracted from maternal blood to be used for screening the fetus for disease.[178] The origin of this fetal DNA is likely to be from breakdown of trophoblast cells in the maternal circulation. The levels of fetal DNA are reported to range between 0.39 and 11.9 per cent of total DNA in the mother's blood, so should be ample to show up any genetic disorders if present. This fetal DNA in maternal blood is very short-lived so we can be confident it belongs to the present pregnancy and is not a residue from a previous pregnancy. This technique is still relatively new, but if it is found to work successfully then we will no longer need to look for intact fetal cells in maternal blood.

Methods available for prenatal diagnosis are, of course, welcomed but none can be totally accurate in their predictions. Mistakes in interpretation can occur. This is an area where trouble is brewing. Increasingly, parents of children born with birth defects that were missed in the screening procedure are taking legal action against the authorities who provided the service. Even the affected children are doing the same, leading to the ironic situation where they are suing for being allowed to be born. What is likely to happen is that the medical profession, wary of possible litigation, will prefer to err on the side of caution in their interpretation of results. Inevitably, some healthy pregnancies will be terminated. The quest for the 'perfect baby' can have undesirable consequences.

The second reason for our interest in these fetal cells is that they can persist in the mother for a very long time—in excess of 30 years has been reported.[179] This is an astonishing finding. It means that, following pregnancy and unbeknown to her, a woman can harbour 'foreign' cells in her body, perhaps for the rest of her life. In other words, she has become a 'chimera', a term adopted from Greek mythology of a fire-breathing creature with the head of a lion, the body of a goat, and a serpent's tail. In modern terminology, a chimera refers to a creature consisting of two or more genetically different kinds of cells or tissues. Since only a small number of these foreign cells are present in the mother, the situation is described as 'microchimerism'. What effects do these fetal cells have on the mother? They are alive and many are also immunocompetent, that is they are capable of immune functions. Will these fetal cells recognize the mother as foreign and attack her in a similar way as a transplant recipient attacks a foreign graft? An analogous situation is seen in bone marrow transplantation. Here, the grafted bone marrow cells can reject the recipient in contrast to the usual situation where it is the recipient that rejects the graft. This rejection in the reverse direction is known as graft versus host disease (GVHD). Does GVHD occur in maternal microchimerism? There is a strong suspicion that fetal cells might be responsible for a disease in the mother called 'scleroderma', which has all the hallmarks of a GVHD. At present, scleroderma is classified as an 'autoimmune' disease, one in which the body's own immune system turns against itself. But if scleroderma is caused by foreign fetal cells attacking the mother, then the disease is obviously no longer 'autoimmune' but is actually 'alloimmune'. There are many other so called 'autoimmune' diseases, such as systemic lupus erythematosis (SLE) and rheumatoid arthritis. Are these also caused by residual fetal cells transferred to the mother during pregnancy? This could explain why 'autoimmune' diseases are found more frequently in adult women following their child-bearing years than in men. The cause of many diseases may need to be reclassified.

Since fetal cells cross the placenta and colonize the mother, we may assume that maternal cells can do the same in the opposite direction. However, the prevailing blood pressure is higher on the fetal side so that cells tend to be more readily transferred from fetus to mother than from mother to fetus. Consequently, fetal cells are found in maternal blood more frequently than maternal cells in fetal blood. As to what these maternal cells might do to the fetus, we can only speculate. In animal experiments maternal cells can attack the fetus resulting in the condition of 'runting', when offspring are born small and stunted with a variety of abnormalities. Runting is a classic GVHD. At the moment, there is no definite evidence that full-blown runting occurs in humans, but there might be less obvious manifestations of this condition which we have not yet recognized. These maternal cells can also remain alive in the fetus in the same way as fetal cells do in the mother. A pregnant mother then can be colonized by cells from her fetus while, at the same time, she herself may also harbour cells from her own mother during her time *in utero*. This double microchimerism makes a mockery of the definition of 'self' as described in Chapter 6. We are not who we think we are! Perhaps each of us should now regard ourselves individually not as 'I' but as 'we'. Might this altered perception lead to a less selfish society?

The effects of alien microchimeric cells may not always be harmful to the host harbouring them. Just because they are frequently observed in association with certain diseases does not necessarily mean they are responsible for causing the disease. They may be just innocent bystanders. These cells could even be beneficial to the host. Microchimeric cells have been seen to replace the functions of deficient bone marrow cells and also help to repair diseased or injured tissues in a variety of organs they have colonized, such as the liver or thyroid. What is interesting is that these fetal cells even begin to look like the surrounding cells in the maternal organs they have in-filtrated. This implies that fetal microchimeric cells are truly stem cells which have the full potential to develop into a multitude of dif-ferent cell types that can replace diseased maternal tissues, a natural example of stem cell therapy (see Chapter 9). The mother may gain from being a chimera.

Surprisingly, even cancer cells, with their aggressive tendency, do not normally succeed in crossing the placenta. Over the past 100 years, no more than a dozen or so probable cases of maternal–fetal transmission have been reported. There are many illustrations in the literature showing masses of cancer cells that have accumulated in the intervillous blood spaces in direct contact with the placenta but have not succeeded in penetrating the layer of syncytiotrophob-last. It seems that maternal cancer, even if it has succeeded in spread-ing to the placenta, is actually prevented from invading any further into the fetal side by the syncytiotrophoblast. Why syncytiotro-phoblast is so resistant is not known. Even if a few cancer cells suc-ceed in penetrating the placental barrier, the fetus' immune defence would probably kill them off before they could establish themselves since they are 'foreign' cells. In 2009, Japanese scientists analysed the genetics of a rare tumour that was transmitted from a mother to her baby.[180] They found the genes coding for the mother's HLA were missing. The reader will remember from Chapter 6 that HLA are the proteins on the surface of our cells that permit us to distinguish 'self' from 'non-self'. So, since the mother's HLA proteins are not present in the tumour, the fetus' immune system can no longer recognize it as foreign and attack it. This particular mutation has enabled the mother's tumour to become immunologically invisible allowing it to survive in the fetus. The syncytiotrophoblast layer of the placenta also employs this useful disguise to evade maternal immune response.

Cancers arising from the fetus also rarely spread to the mother, but there is one conspicuous exception. This is choriocarcinoma, a cancer that we already encountered in Chapter 3. This cancer spreads rapidly into the mother. On anatomical grounds we can see why. Trophoblast lies in direct contact with maternal blood so that cancer arising from trophoblast will shoot straight into the mother's blood stream. The barrier effect of the placenta no longer applies. The maternal organ most frequently affected is her lungs because her lungs are the first port of call for blood returning from the placenta. Trophoblast regularly spreads to the mother's lungs during normal pregnancy too (see Chapter 3).[181]

Tucked in between baby and mother, the placenta keeps a watchful eye over both. Its main concern is to see that the baby gets all it needs from the mother. At the same time, it must ensure nothing harmful comes across. But things do not always go according to plan, as is evident from the variety of problems that can arise. Nature is not infallible. Perhaps a few more millennia of evolution will iron out the glitches. Meanwhile, we should be satisfied with our present placental model which, considering the complexity of its task, is performing with commendable efficiency.

9

Journey's End

The cycle of life and death is inevitable. Birth is when the placenta dies and the baby's life begins. With the baby's first independent breath, the placenta is no longer needed. It has nursed the baby through a potentially perilous intrauterine odyssey. Now the job is done. The clamping of the cord heralds the final parting of the ways. No wonder many cultures today treat the placenta with great respect and even look upon it with genuine affection. It is not just superstition or religious belief. There is a deep sense of loss that, when one life comes into being, another has ended.

At one time, the placenta must have appeared to be a rather mysterious structure, attached to and yet not an integral part of the baby. 'What exactly is it?'; one can almost hear the ancient sages debating the question.[182] Since its physical nature could not be clearly defined, the placenta was accorded a spiritual, even religious, status. In different parts of the world, the relationship of the placenta to the baby varies from being considered a friend (Nepal), an elder sibling (Malaysia), a twin (Nigeria), or part of the baby itself (Hawaii).

According to Anne Fadiman, who studied the Hmong tribe in Laos, the word placenta is translated there to mean a 'jacket' which is considered to be the person's first and finest garment. When a Hmong dies, his or her soul must go back to where the placental 'jacket' is buried and put it on. Only after the soul is properly dressed in the clothing in which it was born can it continue its journey to the 'place beyond the sky' to be reunited with its ancestors. Someday it will be reborn to become the soul of a new baby. If the soul cannot find its 'jacket', it is condemned to an eternity of wandering in the wilderness, naked and alone. A wonderful story of misfortune and redemption.

In some cultures, any harm done to the placenta is thought to induce similar injury or bad luck to the owner's life. This reverence for the placenta is reflected in the respectful ways by which the placenta is disposed of in many traditional societies, unlike the Western world where the organ is generally considered to be just a lump of inanimate flesh which is perfunctorily thrown into the dustbin, incinerated, or used to fertilize rose bushes. The reason why there is so little reverence for the placenta in Western culture may be because the process of birth has become so cold and clinical that it is no longer associated with the beginning of life. The thesis by Sabine Wilms on *Childbirth Customs in Early China* provides an invaluable insight into how the ancient world dealt with the placenta.[183] The proper way to bury the placenta was described in great detail in Chinese medical literature dating back to before the 2nd century BC, and is to some extent is still practised today. Usually, the placenta is buried securely underground. This is to prevent it from being stolen by 'evil spirits' or eaten by wildlife, thereby ensuring that the baby will have a long and healthy life. For example, it is believed that if the placenta is eaten by dogs or pigs, the baby will suffer from manic depression, if eaten by ants the baby will suffer from skin sores, and if eaten by birds the baby may die suddenly. The placenta is always buried face down with the smooth side up. If buried upside down, the baby might vomit during feeding. The ground is chosen as the final resting place because Earth is revered as the creator of all life so it is natural that the placenta should be returned to Her. What better fate is there for the placenta than to emerge from the womb of its natural mother and be immediately engulfed by the enfolding arms of Mother Earth?

In China, the site of burial of the placenta also has great significance. Based on the 'Yin and Yang' dualism of Tao, burying the placenta either in a shady (Yin) or sunny (Yang) side of the wall is believed to enable the parents to choose the gender of the next child. The 'Yang' position facing the sun is usually favoured because this will guarantee the next child will be a boy. Furthermore, a 'Yin' location could weaken the child's 'Chi' and therefore result in poor health. The custom of planting a tree, such as a banana, palm, or sago on top of where the placenta is buried is also a common practice in many parts of the world. The plant emerging from the buried area is then named after the child and must not be cut down, otherwise bad luck will befall the child. In some cultures, instead of being buried in the ground, the placenta is thrown into a river in the belief that 'as the river flows, the child's life will flow with it'.

The use of the placenta as medicine has an extensive history. This can be seen in the *Compendium of Materia Medica* compiled by the Chinese scholar Li Shizhen in 1593, which covers the period between remote antiquity and the Ming Dynasty.[184] Beside recipes for converting placentas into medicinal compounds, this Compendium also contains quaint observations, such as that placentas from first born babies are best in quality and placentas from male and female babies should be used to treat diseases of men and women respectively, which seems logical enough.

The cosmetics industry has long exploited the placenta, using its hormone-rich tissue as a basis for formulating skin creams and other beauty aids. Pharmaceutical companies are now trying to develop products which might be clinically useful. A cut or a wound is invariably surrounded by an area of inflammation. We are all familiar with the symptoms of inflammation: the area is red, hot, swollen, and, above all, painful. At the site of implantation the placenta causes something very similar to a wound on the surface of the uterine lining. Yet, there is no sign of any inflammatory response. The placenta produces a wide variety of compounds, some of which are anti-inflammatory. If these compounds can be developed into therapeutic agents they will be invaluable, particularly for the treatment of longstanding inflammatory conditions such as rheumatoid arthritis, sufferers of which can be totally incapacitated by their swollen and excruciatingly painful joints. Present day treatment is by the administration of steroids, a drug that is not without dangerous side effects, but if this regime can be replaced by a naturally occurring anti-inflammatory compound derived from the placenta it will be a most welcome addition. Interestingly, women with rheumatoid arthritis notice an improvement in their symptoms when they are pregnant.

After inflammation comes the process of healing. The placenta may have something to offer here too. When it peels off from the uterus during delivery, a large raw area is left which quickly heals without formation of scar tissue, unlike in other parts of the body. Again, this could be due to something secreted by the placenta. Every year, large numbers of people undergo surgery of one kind or another. A scar-preventing drug made from the placenta would revolutionize the clinical management of wound healing and skin grafts, especially in cosmetic surgery, a potential billion dollar industry.

The most exciting discovery in recent years is that various components of the placenta are potential sources of stem cells. These immature cells are very much in the news because of their capability to develop into any cell type of choice, thereby providing a means of

216

replacing old or diseased tissues with new ones. The era of 'new lamps for old' is fast becoming reality. A major hurdle is finding a ready supply of stem cells. In 1998, James Thomson from the University of Wisconsin isolated embryonic stem cells from the inner cell mass (ICM) of human blastocysts that were superfluous to requirement for IVF. You may recall from Chapter 3 that the ICM is the population of cells within the blastocyst destined to form the embryo, while the trophectoderm is for making the placenta. Because cells of the ICM are undifferentiated, that is, they are uncommitted as to what kind of cell they will eventually become, they are an ideal source of stem cells. But to harvest these ICM cells for use as stem cells, the embryo has to be sacrificed. If stem cells can be acquired from the placenta, this would solve much of the current ethical dilemma associated with deriving stem cells from embryos.

Blood cells from the umbilical cord have the characteristics of stem cells. While cord blood cells are unlikely to have the full potential to become every type of cell in the body (totipotent) like those derived from embryos, they might, nevertheless, be induced to develop into different kinds of blood cells (pluripotent). This would be particularly useful for the person to whom the placenta belongs should he or she need to replace blood cells in the future. Since these stem cells belong to 'self', they will not be rejected by the recipient's immune system. Many companies have already started to freeze down cord blood to be kept in storage for just such an eventuality. They are expensive, charging around £2000 a time to freeze down a sample for a period of 25 years. There is, of course, no guarantee these cells will still be viable after such a long time preserved in liquid nitrogen. Should one be selfish or altruistic with these stem cells? To be selfish means having the cord blood stored privately only to be used if it is needed by the person to whom the placenta belonged. The altruistic option is to donate the cord blood to a public blood bank to be made freely available to anyone who needs it, although this will lose the benefit of being 'self'. The likelihood of a child needing his or her own stem cells to treat a future disorder is very slim. A potential disease might be leukaemia. This disease originates from a genetic malformation of the white blood cells so that a child's stored stem cells may already contain the same abnormality that caused the child to become ill in the first place. They would be better off using closely-matched stem cells from somebody else rather than their own cells. Altruism, then, should triumph over self-interest.

In addition to cord blood, the umbilical cord itself harbours stem cells. The wall of the cord is made up of a mucous substance called 'Wharton's Jelly', named after Thomas Wharton who first described it in 1656. During early development, large numbers of primitive cells migrate between the embryo and placenta. Many of these become trapped in Wharton's Jelly and are available for harvesting when the placenta is delivered. These cells, known as mesenchymal stem cells, are capable of differentiating into all kinds of tissues and are thus more versatile than cord blood stem cells.

Another potential source of stem cells is the amnion.[185] This is the thin, tissue paper-like membrane which clothes the surface of the placenta. For over 100 years, the amniotic membrane has been used to cover skin wounds, burns, and ulcers and is known to hasten healing so it appears to have regenerative properties. How it works has never been clear. The presence of stem cells within the membrane could now offer a plausible explanation. There are now many reports of cells from the amnion being persuaded to differentiate into a variety of cell types when cultured in the laboratory. These cells will have the potential to replace and regenerate damaged tissues in other organs. Amnion stem cells might not even need to differentiate into the 'correct' cell type to colonize and replace injured areas. Animal experiments have shown that secreted substances from the stem cells can, by themselves, promote repair. This observation is exciting and could lead to an important modification of therapeutic regimes. Stem cells from the amnion could, in future, be cultured in the laboratory and the culture fluid containing all the beneficial products from these cells instead of the cells themselves could then be injected into damaged tissue. This would avoid potential complications associated with the transplantation of cells, such as rejection by the recipient.

Amniotic stem cells are constantly shed into the amniotic cavity throughout pregnancy, so these stem cells can be harvested by amniocentesis during pregnancy in the same way as for obtaining fetal cells for prenatal diagnosis. There is no need to wait until the placenta is delivered. Stem cells in the amniotic cavity already have wound-healing properties. Animal experiments have shown that wounds inside the uterus heal faster than wounds outside. This is a novel function for amniotic fluid. We know it protects the baby by providing a liquid buffer against mechanical shock. That it can also repair intrauterine injuries sustained by the baby is an additional bonus.

With an almost inexhaustible supply, cord blood, umbilical cord, and amnion could all become major sources of stem cells in future. This has added a new dimension to the placenta story. Not only is it the engine that starts intrauterine life, we now see it also helps to sustain life afterwards outside the womb. An International Placenta Stem Cell Society was inaugurated in 2010 to cater for increasing interest in this subject. It is noticeable that many recent research papers in this field have originated from biotech companies rather than from academic institutions. The era of placenta therapeutics is here. What started as scientific endeavour is now firmly established in the marketplace. Unfortunately, whenever money comes into the picture dubious activities will follow. There is already a court case pending of someone accused of harvesting stem cells from cord blood without permission and selling them illegally.

Animals frequently eat their placentas and derive considerable nutritional benefits from doing so since placental tissue is rich in protein. Surprisingly, both carnivores and herbivores do so. There is a theory that this might be a defensive tactic. Leaving the placenta behind would attract the unwanted attention of predators. Better to remove all evidence that a birth has taken place and a defenceless newborn is in the vicinity. This habit is uncommon in humans but does occur sporadically, more in some cultures than in others.[186] Type the word 'placenta' into a search engine and up pops a variety of ingenious recipes of how to cook it! It is not clear why, in contrast to other animal species, humans have largely discarded the habit of 'placentophagy' (eating the placenta). Perhaps the thought that this might be construed as a kind of cannibalism sits uncomfortably in our minds. Animals that eat their placentas do not normally devour their young at the same time. How they distinguish between the two is an interesting question as they are both derived from the same source. Animals, such as mice, do sometimes devour their young. This happens more often with the first litter, suggesting the mother learns not to do so with subsequent pregnancies. Smell seems to be important in this learning process because anosmic mice cannot make the distinction between offspring and placenta. Also in mice, artificial disruption of a gene known as *Mest* will abolish placentophagy proving that the inclination of mice to eat their placentas is, to some extent, genetically programmed. But in humans, the decision must be mostly cultural.

Should all placentas be examined routinely for signs of disease after delivery? Some health authorities recommend that this should be done. This, of course, would be a Herculean task and cannot be taken on lightly for obvious logistical reasons. What would we get out of it? The present consensus among clinicians and pathologists is that there are no real scientific or medical reasons for doing so. Any disease in the placenta will usually be visible either in the baby or the mother or both. The yield of any additional useful information from examining the placenta is too small to justify the time and effort required. There are some cases where it might be useful. For example, in the disease malaria the causative parasites tend to amass in large numbers within the blood spaces of the placenta so seeing them there will confirm the presence of infection, especially when the parasites are difficult to demonstrate elsewhere in the baby or mother.

There is now a new resurgence of interest in routinely recording the size, shape, and weight of the delivered placenta.[187] The stimulus for this has come from studies on fetal programming (see Chapter 5). Research in this area has shown that deviations from the norm for any of these placental parameters taken together with low fetal birth weight reflects a poor intrauterine environment and predicts that the baby will have a high risk of heart disease in adult life. This is an important association that should be documented in birth records. However, since the practice of routine examination of the placenta has been abandoned for so long, it might be difficult to resurrect the habit.

In cases of obstetric litigation, there is a legal requirement to examine the delivered placenta. This usually arises when a baby is born with some kind of birth defect, such as cerebral palsy. The court is asked to rule on whether these defects have resulted from negligence on the part of the attending obstetrician or if there were conditions during pregnancy that might be responsible. The most important cause in utero that can disrupt proper fetal development is lack of oxygen over a prolonged period, a condition known as 'chronic fetal hypoxia'. To the expert eye, certain subtle changes can be discerned in the placenta that might be indicative of this condition, but the interpretation of these findings is likely to come under severe cross-examination by opposing council because not all infants with these changes in the placenta suffer from ill effects. The pathologist who is required to give an opinion should be prepared for a bumpy ride. But analyses of insurance company reports are reassuring. They show that, in most cases when a placenta is available for examination, the verdict is found in favour of the defending obstetrician. Obstetrics, unfortunately, is a specialty particularly prone to litigation. Pregnancy is perceived to be a natural event and any untoward outcome is deemed to be due to medical negligence in the eyes of the public until proven otherwise.

Although not for routine examination, placentas are regularly collected for research purposes. For medical and ethical reasons, detailed studies of the placenta cannot be carried out in its natural state inside the pregnant woman (*in vivo*). Experiments have to be created in the laboratory using cells extracted from the delivered placenta (*in vitro*). These experiments are designed in such a way as to resemble the real life situation as closely as possible. Examining full term placentas delivered at birth allows us a glimpse into what happens at the end of pregnancy. But the more important and interesting events, such as implantation, occur earlier in gestation. This is why researchers prefer to use younger placentas from before the 12th week of pregnancy. These specimens can be obtained from therapeutic terminations performed for social or other reasons. The use of these specimens is governed by strict ethical guidelines by which permission is sought from each patient and total anonymity is maintained. Great emphasis is placed on the fact that only tissues from the placenta will be used for research and never those from the fetus. This distinction is of paramount importance because there are legal as well as emotional implications attached to the use of fetal material for research. The placenta, on the other hand, is considered to be mostly an ancillary appendage which is discarded anyway after each pregnancy and is, therefore, not accorded the same status as the fetus itself. This view is generally appreciated by the ethical committees that govern research as well as by the patients themselves. Permission is usually granted. These are sensitive issues but they are an integral part of doing research on the human placenta. Getting a regular supply of placental tissue is crucial. This is entirely dependent on the goodwill of patients and the cooperation of hospital staff. Otherwise, all projects will grind to a halt. I sometimes feel that the short statement of acknowledgement at the end of every scientific paper on the placenta does not adequately convey the debt we researchers owe to all the people behind the scenes who make our work possible.

The human placenta is so many different things packaged into a single organ that it defies conventional biological classification. Is it a transplant, a cancer, a parasite? It has the hallmarks of all of these. Preceding chapters in this book have detailed the placenta's extraordinary range of functions which are unequalled by any other organ of the body. They are so diverse and intricate that we can only marvel at how the placenta manages to do them all so efficiently. We take it for granted that, normally, all will be well. It is only when the machinery breaks down that we realize how much a successful pregnancy outcome owes to a healthy placenta.

What the placenta is, what it can do, what it tells us about the way we reproduce, and how this knowledge impacts on other areas of human biology should be matters of concern to anyone who is at all scientifically curious about the natural world. Some pregnant women may take a personal interest in the placenta and would be keen to learn more about this remarkable extra-embryonic organ which provides the vital link between herself and her baby. For an organ that is largely ignored during pregnancy and perfunctorily discarded soon after birth, its life is remarkably eventful. The story of the placenta is thoroughly fascinating and deserves to be widely recognized.[188] I have enjoyed telling this story and I hope this enjoyment is shared by those who have read it. 'So brief a life yet so much accomplished' would be a fitting epitaph.

Glossary

Adaptive immunity A highly specialized form of defence against foreign invaders. Utilizes T and B lymphocytes as the immune cells. Comes into play in infections and graft rejection (see innate immunity) .

Aerobic Organisms that require oxygen to survive (see anaerobic).

Allogeneic Individuals of the same species who are genetically different. Almost everyone in the human population is allogeneic to each other (see syngeneic).

Allograft Tissue or organ transplanted between two individuals of the same species who are genetically different (e.g. human kidney, heart, lung transplants).

Altricial Offspring born very immature and need to be nursed (see precocial).

Amino acid A special group of organic compounds that are the building blocks for proteins.

Amniocentesis Withdrawing a sample of amniotic fluid for prenatal diagnosis of chromosomal abnormalities in the embryo.

Amnion A thin membrane which surrounds the embryo and covers the surface of the placenta.

Amniotic cavity Formed when the amnion completely encloses the embryo.

Amniotic fluid Fluid within the amniotic cavity. The embryo is totally immersed in this fluid and is protected by it.

Anaerobic Organisms that live in the absence of oxygen (see aerobic).

Androgenome Artificially created embryo containing only chromosomes from the male (see gynogenome).

Angiogenesis Formation of new blood vessels. An essential process in establishing the circulation in the placenta.

Anosmic Having no sense of smell. Anosmic mice do not eat their placentas.

Antibody A special kind of blood protein produced by B lymphocytes. Antibodies attack foreign entities that have gained access to the body, such as infective organisms, to neutralize them. Also called immunoglobulins. Divided into 5 classes: IgG, IgM, IgA, IgE, and IgD.

Autoimmune disease A disease caused by an individual's immune system attacking his/her own cells or tissues.

Blastocyst The hollow sphere of cells consisting of the inner cell mass and the surrounding trophectoderm that is produced after fertilization of an egg by a sperm. This is the structure that will implant into the uterus.

B lymphocyte A type of cell that produces antibodies in the adaptive immune response (see T lymphocyte).

CAM Cell adhesion molecules. A group of cell surface molecules that promote adhesion between cells or between cells and the extracellular matrix (ECM). An important family are the integrins (see ECM and integrin).

Chimera An animal harbouring 'foreign' cells which are not its own. First described in Greek mythology. When a small number of cells are involved this is known as microchimerism.

Choriocarcinoma A cancer that has arisen from trophoblast.

Chromosomes Pairs of thread-like structures in a cell nucleus that carry the genetic information of the cell. Humans have 46 chromosomes in each cell, 23 of which are inherited from the father and 23 from the mother (see sex chromosomes).

Cloning Technique to create an animal from a single cell. Such an animal is a clone of the individual from which the cell is taken. 'Dolly' the sheep was a clone.

Convergent evolution When two organisms share similar features which have evolved independently.

Corpus luteum Glandular tissue formed in the ovary at the site of a rup-
tured egg follicle after ovulation. Secretes the hormone progesterone (see
ovulation).

CRH Corticotrophin releasing hormone. Another hormone produced by
the placenta. Its function might be to initiate labour.

Cytokine A protein released by a cell which has an effect on another cell,
usually nearby (see hormone).

Decidua This layer lines the cavity of the pregnant uterus. It is transformed
from the non-pregnant endometrium by pregnancy hormones to play an
important role in implantation of the placenta (see endometrium).

Diapause A fertilized egg in a state of suspended animation inside the uterus
waiting for signals to develop further. Used by many animals, such as the
kangaroo.

Differentiated A cell that has already developed its final specialized characteristics. Can no longer be changed into another type of cell. In contrast, an
undifferentiated cell can be induced to become a variety of different kinds
of cells (see stem cell).

DNA Deoxyribonucleic acid. The genetic material which controls heredity.
Located within the chromosomes inside the nucleus. Composed of two
chains that are wound round each other in the famous 'double-helix' described by Watson and Crick. The molecule can make exact copies of itself
when the cell divides, thereby passing on the genetic material to the daugh-
ter cells. Changes in DNA cause mutations (see RNA).

ECM Extra cellular matrix. The ground substance in between cells that bind
cells together like cement in a brick wall (see CAM). There are two main
types: laminin (LM) and fibronectin (FN).

Ectopic pregnancy A pregnancy that occurs outside the uterus, such as in
the Fallopian tube (tubal pregnancy) or in the abdominal cavity (abdominal
pregnancy).

Embryo Very early stage in the formation of the baby inside the womb between 3–8 weeks of pregnancy. After 8 weeks, the embryo becomes a fetus
(see fetus).

Embryology A study of the development of the embryo.

Endocrine glands They produce hormones. The placenta can be considered to belong to this group (see hormones).

Endocrinology The study of hormones.

Endometrium The layer which lines the cavity of the non-pregnant uterus. Its composition changes throughout the menstrual cycle. It breaks down and is shed at menstruation and regenerates at the start of the next cycle. If pregnancy occurs, the endometrium is converted into decidua (see decidua).

Endotheliochorial placenta A moderately invasive type of placenta used by many carnivores.

Endothelium The layer that lines the inside of the heart, blood vessels, and lymphatic vessels (see epithelium).

Endovascular trophoblast Population of trophoblast that invades down the lumen of uterine arteries. Forms an important plug that prevents uterine blood from entering the intervillous space at too high a pressure.

Epidemiology The study of the incidence and prevalence of diseases in different populations. The original idea for fetal programming was based on epidemiological studies.

Epigenetic Modification of gene function without changing the structure of DNA by mutation (see methylation).

Epitheliochorial placenta A non-invasive type of placenta used by many species of domestic animals.

Epithelium The layer that covers the external surface of the body (e.g. skin) and also lines all the hollow structures within the body (e.g. lungs, intestines, bladder etc.) (see endothelium).

Estrogen Hormone that controls female sexual development. Produced by the ovary and also by the placenta.

Estrus When a female animal is 'in heat' and is receptive to mating with a male.

Eutherian Mammals with well-developed placentas that give birth to live offspring. Humans belong to this group.

Fallopian tube A pair of tubes that convey eggs from the ovary to the uterus.

FcRn Neonatal receptor for Fc region. Expressed on the surface of syncytiotrophoblast. Transfers antibodies across the placenta from mother to fetus.

Fetal programming Events occurring in the womb can have profound effects on the development of diseases in adult life.

Fetus A later stage in the formation of the baby inside the womb after 8 weeks of pregnancy (see embryo). The early spelling of 'foetus' is no longer used as it has been shown to be etymologically incorrect.

Genome The complete set of genes in an organism.

Genomic imprinting Gene activity depending on whether it is inherited from the father or mother. An active gene from the father is described as paternal expressed while from the mother is maternal expressed.

GVHD Graft versus host disease—when donor cells attack the recipient instead of the recipient rejecting donor cells.

Gynogenome Artificially created embryo containing only chromosomes from the female (see androgenome).

Haemochorial placenta A deeply invasive type of placenta used by humans and other primates as well as by rodents.

HCG Human chorionic gonadotrophin. The earliest hormone produced by the placenta. Its detection in the woman's urine signals the onset of pregnancy.

HERV Human endogenous retrovirus. Makes syncytin for the formation of syncytiotrophoblast as well as many other functions related to placental development.

Heterozygous Describes an individual in whom members of a pair of genes determining a particular characteristic are dissimilar. If one gene is defective, the other usually remains normal and will compensate for the defective one (see homozygous).

Histology The study of the structure of tissues under a microscope.

HLA Human leukocyte antigen. Cell surface molecules that are highly variable so that almost all individuals will have different HLA on their cells. These molecules are responsible for graft rejection (see MHC).

Homeostasis Maintaining the body's internal physiological systems (e.g. blood pressure, body temperature) at a constant level.

Homozygous Describes an individual in whom both members of a pair of genes determining a particular characteristic are identical. If one gene is defective, then the other will be also. This can lead to 'recessive' diseases, such as cystic fibrosis (see heterozygous).

Hormone A substance produced by certain tissues of the body (endocrine glands) which is carried in the blood stream to affect the function of tissues at a distance from the source of production. Placental hormones are good examples (see cytokine).

HPL Human placental lactogen. Another hormone produced by the placenta. Prepares for breast milk production in the mother.

Hybrid Progeny of a cross between two different species.

Hydatidiform mole Placenta looks like a bunch of grapes. Paternal genes only are present. It is a naturally occurring androgenome.

Hypothalamus A region at the base of the forebrain. It is an important centre for control of homeostasis, sexual activity, and maternal nurturing of the young. Responds to placental hormones.

ICM Inner cell mass. The group of cells inside the blastocyst that will develop into the embryo.

Implantation Attachment of the early embryo to the lining of the uterus followed by invasion of the underlying uterine tissue by trophoblast. This is the beginning of the formation of the placenta.

Imprinted genes These genes are either paternal or maternal expressed (see genomic imprinting). Many imprinted genes are found in the placenta.

Innate immunity The body's first line of defence against infections. Utilizes a variety of cells for its defence activity, including NK (Natural Killer) cells (see adaptive immunity).

Integrin A family of cell surface molecules that act as receptors for ECM, thereby permitting cells to adhere (see CAM and ECM).

Interstitial trophoblast Population of trophoblast that invades into decidua to remodel the uterine arteries.

Intervillous space Area where the placenta is bathed in a pool of maternal blood created by syncytiotrophoblast destroying uterine blood vessels during implantation. This is typical of haemochorial placentation.

In vitro Biological processes occurring in cells or tissues grown in culture in the laboratory (see *in vivo*).

In vivo Biological processes occurring in a living organism (see *in vitro*).

IUGR Intrauterine growth restriction. Due to inadequate nutrition inside the womb caused by poor trophoblast remodelling of uterine arteries, a baby is born smaller than normal. It will be susceptible to heart and other diseases in adult life.

IVF *In vitro* fertilization. Fertilization of an egg by a sperm outside the body. The fertilized egg is then put back into the uterus for further development.

KIR Killer-cell immunoglobulin-like receptor. Expressed on the surface of NK cells. Recognizes HLA molecules on target cells which either activate or inhibit the action of the NK cells. An important receptor for maternal recognition of the placenta (see NK cell).

Knock out (KO) A useful technique to target and destroy a single gene in experimental animals. Used to find out the function of a gene.

LH Lutenizing hormone. This is a gonadotrophin secreted by the anterior pituitary gland at the base of the brain. Stimulates ovulation.

Marsupial A group of mammals with poorly developed placentas whose offspring are born in a very immature state and have to be nursed in a pouch (marsupium). The kangaroo is a good example.

Methylation Addition of a methyl chemical group to DNA resulting in a change in gene function without alteration of the structure of the DNA (see epigenetic).

MHC Major Histocompatibility Complex. A group of genes located on chromosome 6 in humans that code for many important molecules, including HLA (see HLA).

Mitochondria Tiny structures in the cytoplasm of every cell that produce energy for the cell by aerobic respiration using oxygen. Interestingly, mitochondria are inherited from the mother.

MMP Matrix-degrading metalloproteinase. An enzyme that digests ECM to facilitate trophoblast invasion (see ECM).

Monotreme Primitive mammals that lay eggs instead of giving birth to live offspring. The only two living species are the duck-billed platypus and the spiny anteater.

Myometrium The muscle layer of the uterus. It lies underneath the endometrium.

Neutrophil A type of white blood cell capable of ingesting and killing bacteria. Provides important defence against infections. The process is known as 'phagocytosis' (see phagocyte).

NK cells Natural Killer cells. They are a component of the innate immune system. They can kill certain targets, such as virus-infected or tumour cells. In the context of pregnancy, they are the cells by which the mother interacts with the placenta.

Oocyte A progenitor cell in the ovary that will mature into an ovum (egg).

Oviparity Laying eggs (see viviparity).

Ovulation Release of an egg by the ovary. Normally occurs on the 14th day of the menstrual cycle. Stimulated by lutenizing hormone (LH) secreted by the anterior pituitary (see LH and corpus luteum).

Parthenogenesis Self-fertilization of an egg without a sperm.

PE Pre-eclampsia. A serious disorder of pregnancy diagnosed by a sudden rise in blood pressure in a pregnant woman. Poor trophoblast remodelling of uterine arteries with resulting inadequate nutrition for the fetus and placenta could be a primary cause of the disease.

Peptides Chains of amino acid molecules linked together.

Phagocyte A cell that is able to ingest and digest foreign particles (see neutrophil).

Placenta creta When the placenta adheres too tightly to the wall of the uterus and does not separate during delivery. The condition is usually caused by absence of decidua.

Placenta praevia When the placenta implants on or too close to the opening of the cervix and prevents delivery of the baby.

Placentophagy The practice of eating the placenta.

Platelets Tiny cells in the blood involved in blood coagulation.

Pluripotent Cells capable of developing into many different cell types but not all (see totipotent).

Polymorphic Molecules that exist in two or more different forms. HLA molecules are highly polymorphic (see HLA).

Precocial Offspring born already capable of fending for itself (see altricial).

Progesterone A hormone produced by a woman after ovulation during each menstrual cycle. It prepares the uterine lining for pregnancy. The placenta then takes over the production of progesterone.

RNA Ribonucleic acid. It translates the genetic information encoded in DNA to make proteins for the cell (see DNA).

Sex chromosomes These determine the sex of the individual. Women have two X chromosomes (XX), one inherited from the mother and the other from the father. Men have one X and one Y chromosome (XY), the X being from the mother and the Y from the father. To obtain a boy, the mother's egg has to be fertilized by a Y-bearing sperm from the father.

Signal transduction Conversion of the signals produced on the surface of a cell by interaction between integrin and ECM into biochemical signals inside the cell. This makes the cell respond.

Stem cell An undifferentiated cell that can give rise to different cell types and is capable of unlimited cell division. It therefore can be used to replace worn out or injured tissues.

Syncytiotrophoblast The layer of trophoblast that has no intercellular boundaries. It lines the intervillous space and is in contact with maternal blood. It has important protective, transport, and productive functions.

Syngeneic Individuals who are genetically identical (e.g. some identical twins) (see allogeneic).

Teratoma A tumour composed of a variety of poorly differentiated embryonic tissues. Teratoma of the ovary is an example of a gynogenome.

TIMP Tissue inhibitor of MMP. As its name implies, it counteracts the action of MMP (see MMP).

T lymphocyte A special type of cell with many subpopulations that have important roles to play in the adaptive immune response (see B lymphocyte).

Totipotent Cells capable of developing into all cell types (see pluripotent). In mammals the only truly totipotent cells are those in the early embryo.

Trimester The nine months of human pregnancy are divided into three sections termed first, second, and third trimester.

Trophectoderm The layer of cells surrounding the blastocyst that will give rise to trophoblast (see blastocyst).

Trophoblast Cells derived from the layer of trophectoderm lining the blastocyst. Eventually these will form three main populations in the placenta: syncytiotrophoblast, endovascular trophoblast, and interstitial trophoblast.

UPD Uniparental disomy. This is when a pair of chromosomes is inherited from one parent instead of one from each parent. If the pair comes from the father, it is called paternal UPD; if from the mother it is maternal UPD.

Villi Small projections formed on the surface of the placenta. They come into contact with maternal blood in the intervillous space and provide an extensive area for absorption of nutrients and gaseous interchange between mother and fetus.

Viviparity Giving birth to live offspring (see oviparity).

Zoonosis A disease of animals that can be transmitted to humans.

Endnotes

CHAPTER I

1. Hilary Gilson (2008). Leonardo da Vinci's Embryological Drawings of the Fetus, *Embryo Project Encyclopedia*, Arizona State University.
2. Moffett A., Loke C. (2006). 'Immunology of Placentation in Eutherian Mammals', *Nature Review Immunology*, 6: 584–94.
3. Loke Y.W. (1982). 'Transmission of Parasites Across the Placenta', in Lumsden W.H.R., Baker J.R., Muller R. (eds), *Advances in Parasitology*, 21: 155–228.
4. Brambell F.W.R. (1966). 'The Transmission of Immunity from Mother to Young and the Catabolism of Immunoglobins', *Lancet*, II: 1087–93.
5. Nelson J.L. (2012). 'The Otherness of Self: Microchimerism in Health and Disease', *Trends Immunology*, 33: 421–7.
6. Evain-Brion D., Malassine A. (2003). 'Human Placenta as an Endocrine Organ', *Growth Hormone & IGF Research*, 13: S34–7.
7. Keverne E.B. (2006). 'Trophoblast Regulation of Maternal Endocrine Function and Behaviour', in Moffett A., Loke C., McLaren A. (eds), *Biology and Pathology of Trophoblast*, Cambridge University Press, 148–68.
8. Keverne E.B. (2001). 'Genomic Imprinting, Maternal Care and Brain Evolution', *Hormones and Behavior*, 40: 146–55.
9. Johnson M.H. (2007). *Essential Reproduction*, Blackwell, Chapter 10.
10. Moore T., Haig D. (1991). 'Genomic Imprinting in Mammalian Development: A Parental Tug-of-war', *Trends in Genetics*, 7: 45–9.
11. Pizzi M., Fassan M., Cimino M., Zanardo V., Chiarelli S. (2012). 'Realdo Colombo's *De Re Anatomica*: The Renaissance Origin of the Term "Placenta" and its Historical Background', *Placenta*, 33: 655–7.
12. Wooding F.B.P., Burton G.J. (2008). *Comparative Placentation*, Springer-Verlag.

13. Pijnenborg R., Robertson W.B., Brosens I., Dixon G. (1981). 'Trophoblast Invasion and the Establishment of Haemochorial Placentation in Man and Laboratory Animals', *Placenta*, 2: 71–91.
14. King A. (1995). 'Decidualisation', in Loke Y.W., King A. (eds), *Human Implantation: Cell Biology and Immunology*, Cambridge University Press, 18–31.
15. Fox D. (1999). 'Why We Don't Lay Eggs', *New Scientist*, 162: 27–31.
16. Brown S. (2008). 'Top Billing for Platypus at End of Evolution Tree', *Nature*, 453: 138–9.
17. Renfree M.B., Shaw G. (1996). 'Reproduction of a Marsupial: From Uterus to Pouch', *Animal Reproduction Science*, 42: 393–404.

CHAPTER 2

18. Warren W.C. et al. (2012). 'Genome Analysis of the Platypus Reveals Unique Signatures of Evolution', *Nature*, 453: 175–83.
19. Freyer C., Zeller U., Renfree M.B. (2003). 'The Marsupial Placenta: A Phylogenetic Analysis', *Journal of Experimental Zoology*, 299: 59–77.
20. Renfree M.B. (1981). 'Marsupials: Alternative Mammals', *Nature*, 293: 100–1.
21. Steven D.H. (1983). 'Interspecies Differences in the Structure and Function of Trophoblast', in Loke Y.W., Whyte A. (eds), *Biology of Trophoblast*, Elsevier, 111–36.
22. Cockburn K., Rossant J. (2010). 'Making the Blastocyst: Lessons from the Mouse', *The Journal of Clinical Investigation*, 120: 995–1003.
    Rossant J., Cross J.C. (2001). 'Placental Development: Lessons from Mouse Mutants', *Nature Reviews Genetics*, 2: 538–48.
23. Mess A., Carter A.M. (2007). 'Evolution of the Placenta During the Early Radiation of Placental Mammals', *Comparative Biochemistry and Physiology—Part A*, 148: 764–79.
24. Martin R.D. (2008). 'Evolution of Placentation in Primates: Implications of Mammalian Phylogeny', *Journal of Evolutionary Biology*, 35: 125–45.
25. Vogel P. (2008). 'The Current Molecular Phylogeny of Eutherian Mammals Challenges Previous Interpretations of Placental Evolution', *Placenta*, 26: 591–6.
    Wildman D.E., Chen C., Erez O., Grossman L.I., Goodman M., Romero R. (2006). 'Evolution of the Mammalian Placenta Revealed by Phylogenetic Analysis', *Proceedings of the National Academy of Sciences (PNAS) USA*, 103: 3203–8.

26. Enders A.C., Carter A.M. (2012). 'The Evolving Placenta: Convergent Evolution of Variations in the Endotheliochorial Relationship', *Placenta*, 33: 319–26. Endotheliochoral placentas occur in all four major clades of eutherian mammals. Species with this type of placenta include one of the smallest (pygmy shrew) and the largest (African elephant) land mammals.

   Enders A.C., Carter A.M. (2012). 'The Evolving Placenta: Different Developmental Paths to a Haemochorial Relationship', *Placenta*, 33 (Supplement A): S92–8.

   Elliott M.G., Crespi B.J. (2009). 'Phylogenetic Evidence for Early Haemochorial Placentation in Eutheria', *Placenta*, 30: 923–1004.

27. Zhe X.L., Chong X.Y., Qing J.M., Qiang J. (2011). 'A Jurassic Eutherian Mammal and Divergence of Marsupials and Placentals', *Nature*, 476: 442–5.

28. Cloke B., Fusi L., Brosens J. (2010). 'Decidualisation', in Pijnenborg R., Brosens J., Romero R. (eds), *Placental Bed Disorders*, Cambridge University Press, 29–40.

29. Finn C.A. (1996). 'Why do Women Menstruate? Historical and Evolutionary Review', *European Journal of Obstetrics & Gynaecology and Reproductive Biology*, 70: 3–8.

   Finn C.A. (1998). 'Menstruation: A Nonadaptive Consequence of Uterine Evolution', *The Quarterly Review of Biology*, 73: 163–73.

30. Khong T.Y., Robertson W.B. (1987). 'Placenta Creta and Placenta Praevia Creta', *Placenta*, 8: 399–409.

31. Nama V., Wilcock F. (2011). 'Caesarian Section on Maternal Request: Is Justification Necessary?' *The Obstetrician & Gynaecologist*, 13: 263–9. In 2001, in the UK, there were 595,000 live births. Of these, 8747 (1.4 per cent) had caesarean sections on request. Tocophobia (intense fear of childbirth) was the most important reason for request.

CHAPTER 3

32. Moffett A., Loke C., McLaren A. (eds) (2006). *Biology and Pathology of Trophoblast*, Cambridge University Press. This volume is the proceedings of the Novartis Foundation Symposium. The discussions at the end of each chapter are particularly interesting.

33. Burton G.J. (2012). 'The Centre for Trophoblast Research: Improving Health Through Placental Research', *Reproductive BioMedicine Online*, 25: 2–4. Graham Burton has been director of the Centre for Trophoblast Research (CTR) from its formation in 2007.

34. Cross J. (2006). 'Trophoblast Cell Fate Specification', in Moffett A., Loke C., McLaren A. (eds), *Biology and Pathology of Trophoblast*, Cambridge University Press, 3–14.

35. Hubrecht A.A.W. (1889). 'Studies in Mammalian Embryology I, The Placentation of *Erinaceus Europacus*, with remarks on the Phylogeny of the Placenta', *Quarterly Journal of Microscopial Science*, 30: 284–404.

36. Loke Y.W. (1983). 'Human Trophoblast in Culture', in Loke Y.W., Whyte A. (eds), *Biology of Trophoblast*, Elsevier, 663–701.

    King A., Thomas L., Bischof P. (2000). 'Cell Culture Models of Trophoblast II, Trophoblast Cell Lines', *Placenta*, 21, Suppl. A: S113–19.

37. Tanaka S., Kunath T., Hadjantonakis A.K., Nagy A., Rossant J. (1998). 'Promotion of Trophoblast Stem Cell Proliferation by FGF4', *Science*, 282: 2072–5.

    Erlbacher A., Price K.A., Glimsher L.H. (2004). 'Maintenance of Mouse Trophoblast Stem Cell Proliferation by TGF-beta/activin', *Developmental Biology*, 275: 158–69.

    These two papers describe the derivation and maintenance of a stem cell population from early trophoblast cells of the blastocyst stage of the mouse embryo.

38. Hemberger M., Udayashankar R., Tesah P., Moore H., Burton G.J. (2010). 'ELF5-enforced Transcriptional Networks Define an Epigenetically Regulated Trophoblast Stem Cell Compartment in the Human Placenta', *Human Molecular Genetics*, 19: 2456–67.

    Genbacev O., Donne M., Kapidzic M., Gormley M., Lamb J., Gilmore J. et al. (2011). 'Establishment of Human Trophoblast Progenitor Cell Lines from the Chorion', *Stem Cells*, 29: 1427–36.

    These two papers describe attempts to derive human trophoblast stem cells.

39. Hertz R. (1959). 'Chorioncarcinoma of Women Maintained in Serial Passage in Hamster and Rat', *PNAS USA*, 102: 77–80.

40. Pattillo R.A., Gey G.O. (1968). 'The Establishment of a Cell Line of Human Hormone-synthesizing Trophoblast Cells In Vitro', *Cancer Research*, 28: 1231–6.

41. Park W.W. (1971). *Chorioncarcinoma: A Study of its Pathology*, William Heinemann.

    Bower M., Brock C., Fisher R.A., Newlands E.S.N., Rustin G.J.S. (1995). 'Gestational Chorioncarcinoma', *Annals of Oncology*, 6: 503–8.

42. Hawes, C.S., Suskin H.A., Petropoulos A., Latham S.E., Mueller U.W. (1994). 'A Morphologic Study of Trophoblast Isolated from Peripheral Blood of Pregnant Women', *American Journal of Obstetrics & Gynecology*, 170: 1297–300.

43. Knight M., Redman C.W., Linton E.A., Sargent I.L. (1998). 'Shedding of Syncytiotrophoblast Microvilli into the Maternal Circulation in Pre-eclamptic Pregnancies', *British Journal of Obstetrics & Gynaecology*, 105: 632–40.

44. Loke Y.W., Butterworth B.H. (1987). 'Heterogeneity of Human Trophoblast Populations', in Gill T.J., Wegmann T.G. (eds), *Immunoregulation and Fetal Survival*, Oxford University Press, 197–209.

Loke Y.W., King A. (1995). 'Early Trophoblast Development', in Loke Y.W., King A. (eds), *Human Implantation: Cell Biology and Immunology*, Cambridge University Press, 32–45.

Includes many histological pictures, some in colour. Easy to visualize the different trophoblast populations.

45. Dearden L., Ockleford C.D. (1983). 'Structure of Human Trophoblast: Correlation with Function', in Loke Y.W., Whyte A. (eds), *Biology of Trophoblast*, Elsevier, 69–110.

46. Kalter S.S. (1983). 'Viral Expression in the Trophoblast', in Loke Y.W., Whyte A. (eds), *Biology of Trophoblast*, Elsevier, 627–62.

47. Thiry L., Loke Y.W., Whyte A., Hard R.C., Sprecher-Goldberger S., Buekens P. (1981). 'Heterologous Antiserum to Human Syncytiotrophoblast Membrame is Cytotoxic to Retrovirus-producing Cells and to some Cancer Cell Lines', *American Journal of Reproductive Immunology*, 1: 240–5.

48. Muir A., Lever A., Moffett A. (2004). 'Expression and Functions of Human Endogenous Retroviruses in the Placenta: An Update', *Trophoblast Research*, 18: S16–25.

Hayward M.D., Pötgens A.J., Drewlo S., Kaufmann P., Rasko J.E.J. (2007). 'Distribution of Human Endogenus Retrovirus Type W Receptor in Normal Human Villous Placenta', *Pathology*, 39: 406–12.

49. Muir A., Lever A.M., Moffett A. (2006). 'Human Endogenus Retrovirus-W Envelope (Syncytin) is Expressed in both Villous and Extravillous Trophoblast Populations', *Journal of General Virology*, 87: 2067–71.

Holder B.S., Tower C.L., Abrahams V.M., Aplin J.D. (2012). 'Syncytin 1 in the Human Placenta', *Placenta*, 33: 460–6.

Blaise S., de Parseval N., Bénit L., Heidmann T. (2003). 'Genomewide Screening for Fusogenic Human Endogenous Retrovirus Envelopes

Identifies Syncytin 2, a Gene Conserved on Primate Evolution', *PNAS USA*, 100: 13013–18.

50. Macfarlan T.S., Gifford W.D., Driscoll S., Lettieri K., Rowe H.M., Bonanomi D. et al. (2012). 'Embryonic Stem Cell Potency Fluctates with Endogenous Retroviral Activity', *Nature*, 487: 57–63. Many genes expressed during early mouse development are from retroviral elements called MERV (mouse equivalent of human HERV). These genes regulate the production of cell types, for example formation of various placental tissues. Therefore, retroviral genes have an important role to play in early mammalian development besides making syncytin for formation of syncytiotrophoblast.

CHAPTER 4

51. Haig D. (1993). 'Genetic Conflicts in Human Pregnancy', *Quarterly Review of Biology*, 68: 495–532.

52. Lefebvre L. (2012). 'The Placental Imprintome and Imprinted Gene Function in the Trophoblast Glycogen Cell Lineage', *Reproductive BioMedicine Online*, 25: 44–57. Discusses the technical difficulties in determining the imprinting status of placental genes because of the close mixing of fetal and maternal cells. Some genes which were originally designated as imprinted are now shown not to be.

53. Surani M.A., Barton S.C., Norris M.L. (1984). 'Development of Reconstituted Mouse Eggs Suggests Imprinting of the Genome During Gametogenesis', *Nature*, 308: 548–50.

    Barton S.C., Surani M.A., Norris M.L. (1984). 'Role of Paternal and Maternal Genomes in Mouse Development', *Nature*, 311: 374–6.

    Mcgrath J., Salter D. (1984). 'Completion of Mouse Embryogenesis Requires both the Maternal and Paternal Genomes', *Cell*, 37: 179–83.

54. Jones S. (2002). 'Y: *The Descent of Man*', Abacus. Presents the male sex as a threatened species.

    'Steve Jones is much harder on man than I am,' Germaine Greer. 'Surplus to Requirements?' *The Guardian*, 16 November 2002.

55. Moore T. (2012). 'Parent-offspring Conflict and the Control of Placental Function', *Placenta*, 26: Suppl. S33–6.

56. John R.M., Surani M.A. (2000). 'Genomic Imprinting, Mammalian Evolution, and the Mystery of Egg-laying Mammals', *Cell*, 10: 585–8.

    Renfree M.B., Ager E.I., Shaw G., Pask A.J. (2008). 'Genomic Imprinting in Marsupial Placentation', *Reproduction*, 136: 523–31.

57. DeChiara T.M., Robertson E.J., Efstratiadis A. (1991). 'Parental Imprinting of the Mouse Insulin-like Growth Factor II Gene', *Cell*, 64: 849–59.

58. Suzuki S., Renfree M.B., Pask A.J., Shaw G., Kobayashi S., Khoda T. et al. (2005). 'Genomic Imprinting of IGF2, p57 (KIP2) and PEG1/MEST in a Marsupial, the Tammar Wallaby', *Mechanisms of Development*, 122: 213–22. Indicates imprinting of these genes preceded the separation of eutherians from marsupials.

59. Eggenschwiler J., Ludwig T., Fisher P., Leighton P.A., Tilghman S.M., Efstratiadis A. (1997). 'Mouse Mutant Embryos Overexpressing IGF-II Exhibit Phenotype Features of the Beckwith–Wiedemann and Simpson–Golabi–Behmel Syndromes', *Genes Development*, 11: 3128–42.

60. Constância M., Hemberger M., Hughes J., Dean W., Ferguson-Smith A., Fundel R. et al. (2002). 'Placental IGF-II is a Major Modulator of Placental and Fetal Growth', *Nature*, 417: 945–8.

    Constância M., Angiolini E., Sandovici I., Smith P., Smith R., Kelsey G. et al. (2005). 'Adaptation of Nutrient Supply to Fetal Demand in the Mouse Involves Interaction between the IGF2 Gene and Placenta Transporter Systems', *PNAS USA*, 102: 19219–24.

61. Maher E.R., Reik W. (2000). 'Beckwith–Wiedemann Syndrome: Imprinting in Clusters Revisited', *Journal of Clinical Investigation*, 105: 247–52.

    Weksberg R., Shuman C., Beckwith J.B. (2010). Beckwith–Wiedemann Syndrome', *European Journal of Human Genetics*, 18: 8–14.

    These two papers confirm that biallelic IGF2 expression is associated with BWS in humans.

62. Lyon M.F. (1961). 'Gene Action in the X-chromosome of the Mouse (*Mus musculus* L)' *Nature*, 190: 372–3.

63. Heard E., Clerc P., Avner P. (1997). 'X-chromosome Inactivation in Mammals', *Annual Review of Genetics*, 31: 571–610.

64. Takagi N., Sasaki M. (1975). 'Preferential Inactivation of the Paternally Derived X-chromosome in the Extraembryonic Membranes of the Mouse', *Nature*, 256: 640–2.

65. Hiby S.E., Lough M., Keverne E.B., Surani M.A., Loke Y.W., King A. (2001). 'Paternal Monoallelic Expression of PEG3 in the Human Placenta', *Human Molecular Genetics*, 10: 1093–100.

66. Li L., Keverne E.B., Aparicio S.A., Ishino F., Barton S.C., Surani M.A. (1999). 'Regulation of Maternal Behaviour and Offspring Growth by Paternally Expressed PEG3', *Science*, 284: 330–3.

67. Broad K.D., Keverne E.B. (2011). 'Placental Protection of the Fetal Brain during Short-term Food Deprivation', *PNAS USA*, 108: 15237–41.

68. Kajii T., Ohama K. (1977). 'Androgenetic Origin of Hydatidiform Mole', *Nature*, 268: 633–4.

69. Jacobs P.A., Wilson C.M., Sprenkle J.A., Rosenshein N.B., Migeon B.R. (1980). 'Mechanism of Origin of Complete Hydatidiform Moles', *Nature*, 268: 714–16.

70. Loke Y.W. (1969). 'Sex Chromatin of Hydatidiform Moles', *Journal of Medical Genetics*, 6: 22–5.

71. Bracken M.B. (1987). 'Incidence and Aetiology of Hydatidiform Mole: An Epidemiological Review', *British Journal of Obstetrics & Gynaecology*, 94: 1123–35.

72. Fisher R.A., Hodges M.D., Newlands E.S. (2004). 'Familial Recurrent Hydatidiform Mole: A Review', *Journal of Reproductive Medicine*, 49: 595–601.

73. Wu F.Y. (1973). 'Recurrent Hydatidiform Mole: A Case Report of Nine Consecutive Molar Pregnancies', *Obstetrics & Gynaecology*, 41: 200–4.

74. Fallahian M. (2003). 'Familial Gestational Trophoblast Disease', *Placenta*, 24: 797–9.

75. Parrington J.M., West L.F., Povey S. (1984). 'The Original of Ovarian Teratomas', *Journal of Medical Genetics*, 21: 4–12.

76. Francis R.C. (2011). *Epigenetics: The Ultimate Mystery of Inheritance*, W.W. Norton & Co.

77. Manipalviratn S., DeCherney A., Segars J. (2009). 'Imprinting Disorders and Assisted Reproductive Technology', *Fertility & Sterility*, 91: 305–15.

   Grace K.S., Sinclair K.D. (2009). 'Assisted Reproductive Technology, Epigenetics and Long-term Health: A Developmental Time Bomb Still Ticking', *Seminars in Reproductive Medicine*, 27: 409–16.

   Wilkins-Haug L. (2009). 'Epigenetics and Assisted Reproduction', *Current Opinion in Obstetrics & Gynecology*, 21: 201–6.

   These authors questioned the safety of ART since several imprinting disorders have occurred.

78. Katari S., Turan N., Bibikova M., Erinle O., Chalian R., Foster M. et al. (2009). 'DNA Methylation and Gene Expression Differences in Children Conceived In Vitro or In Vivo', *Human Molecular Genetics*, 10: 3769–78. Shows lower levels of DNA methylation in placenta in ART compared to normally conceived babies. Affects both imprinted and non-imprinted genes.

241

CHAPTER 5

79. Burton G.J., Watson A.L., Hempstock J., Skepper J.N., Jauniaux E. (2002). 'Uterine Glands Provide Histiotrophic Nutrition for the Human Fetus During the First Trimester of Pregnancy', *Journal of Clinical Endocrinology & Metabolism*, 87: 2954–9.

80. Hertig A.T., Rock J. (1951). 'The Implantation and Early Human Development of the Human Ovum', *American Journal of Obstetrics & Gynecology*, 61: 8–14.

    Hamilton W.J., Boyd J.D. (1960). 'Development of the Human Placenta in the First Three Months of Gestation', *Journal of Anatomy*, 94: 297–328.

81. Loke Y.W., King A. (1995). 'Surface Epithelium', in Loke Y.W., King A. (eds), *Human Implantation: Cell Biology and Immunology*, Cambridge University Press, 22–7.

    Denker H.W. (1990). 'Trophoblast–Endometrial Interactions at Embryo Implantation: A Cell Biological Paradox', *Trophoblast Research*, 4: 3–29.

82. Rogers P. (1993). 'Uterine Receptivity', in Trounson A., Gardner D.K. (eds), *Handbook of In Vitro Fertilisation*, CRC Press, 263–85.

83. Gerris J., Van Royen E. (2000). 'Avoiding Multiple Pregnancies in ART. A Plea for Single Embryo Transfer', *Human Reproduction*, 15: 1884–8.

84. Galen A., O'Connor J.E., Valbuena D., Herrer R., Remohí J., Pampfer S. et al. (2000). 'The Human Blastocyst Regulates Endometrial Epithelial Apoptosis in Embryonic Adhesion', *Biology of Reproduction*, 63: 430–9. Apoptosis (programmed cell death) induced by the implantating blastocyst on the uterine epithelium is necessary for trophoblast invasion.

85. Ramsey E.M. (1954). 'Circulation in the Maternal Placenta of Primates', *American Journal of Obstetrics & Gynecology*, 67: 1–14.

86. Burton G.J., Hempstock J., Jauniaux E. (2003). 'Oxygen, Early Embryonic Metabolism and Free Radical-mediated Embryopathies', *Reproductive BioMedicine Online*, 6: 84–96. Human placenta is not truly haemochorial until the end of the first trimester.

87. Falkowski P.G., Katz M.E., Milligan A.J. et al. (2005). 'The Rise of Oxygen Over the Past 205 Million Years and the Evolution of Large Placental Mammals', *Science*, 309: 2202–4.

88. Pijnenborg R. (1996). 'The Placental Bed', *Hypertension in Pregnancy*, 15: 7–23.

89. Pijnenborg R., Vercruysse L., Hanssens M. (2006). 'The Uterine Spiral Arteries in Human Pregnancy: Facts and Controversies', *Placenta*, 27:

939–58. The definitive paper of the placental bed based on a lifetime's study.

90. Ramsey E.M. (1962). 'Circulation in the Intervillous Space of the Primate Placenta', *American Journal of Obstetrics & Gynecology*, 84: 1649–63.

91. King A., Loke Y.W. (1994). 'Unexplained Fetal Growth Retardation: What is the Cause?', *Archive of Disease in Childhood*, 70: F225–7.

92. Pijnenborg R., Brosens I., Romero R. (eds) (2010). *Placental Bed Disorders: Basic Science and its Translation to Obstetrics*, Cambridge University Press. The editors of this book were the pioneers who established the importance of trophoblast remodelling of decidual arteries during implantation. They assembled a group of specialists in this field as contributors. Contains a vast amount of valuable information collected in one volume.

93. Redman C.W., Sargent I.L. (2005). 'Latest Advances in Understanding Pre-eclampsia', *Science*, 308: 1592–4.
    Steegers E.A., von Dadelszen P., Duvekot J.J., Pijnenborg R. (2010). 'Pre-eclampsia', *Lancet*, 376: 631–44.

94. Redman C.W. Sacks G.P., Sargent I.L. (1999). 'Pre-eclampsia: An Excessive Maternal Inflammatory Response to Pregnancy', *American Journal of Obstetrics & Gynecology*, 180: 499–506.

95. National Collaborating Centre for Women's and Children's Health (2010). 'Hypertension in Pregnancy: The Management of Hypertensive Disorder During Pregnancy', *NICE Clinical Guidelines*, 107. Recommends high-risk women to take 75mg aspirin daily from 12 weeks of pregnancy until delivery, but this type of prevention is not yet globally accepted.
    National Institute for Health and Clinical Excellence (NICE) (2010). 'Antenatal Care: Routine Care for the Healthy Pregnant Woman', *NICE Clinical Guidelines*, 62. Recommends routine screening for risk factors for pre-eclampsia.

96. Alberry M., Bills V., Soothill P. (2011). 'Review: An Update on Pre-eclampsia Prediction Research', *The Obstetrician & Gynaecologist*, 13: 79–85.

97. Barker D.J. (1990). 'The Fetal and Infant Origins of Adult Disease', *British Medical Journal*, 301: 1111.
    Barker D.J. (2007). 'The Origins of the Developmental Origins Theory', *Journal of Internal Medicine*, 261: 412–17.

98. Thornburg K.L., O'Tierney P.F., Louey S. (2010). 'Review: The Placenta is a Programming Agent for Cardiovascular Disease', *Placenta*, 31: S54–9.

99. Burton G.J., Jauniaux E., Charnock-Jones D.S. (2010). 'The Influence of the Intrauterine Environment on Human Placental Development', *International Journal of Developmental Biology*, 54: 303–12.

Burton G.J., Barker D.J.P., Moffett A., Thornburg K. (eds) (2011). *The Placenta and Human Developmental Programming*, Cambridge University Press.

This was the first time that placentologists and epidemiologists organized a scientific symposium together. The collection of essays in this volume contains the most recent information on this promising area of research.

100. Sandovici I., Hoelle K., Angiolini E., Constância M. (2012). 'Placental Adaptations to the Maternal Fetal Environment: Implications for Fetal Growth and Developmental Programming', *Reproductive BioMedicine Online*, 25: 68–89. A comprehensive up-to-date review.

101. Paul A.M. (2010). 'How the First Nine Months Shape the Rest of Your Life: The New Science of Fetal Origins', *Time Magazine*, October 44–9. Cover story in *Time Magazine*.

102. Maccani M.A., Marsit C.J. (2009). 'Epigenetics in the Placenta', *American Journal of Reproductive Immunology*, 62: 78–89. A variety of placental pathologies are the result of aberrant epigenetic regulation.

Nelissen E.C., van Montfoort A.P., Dumoulin J.C., Evers J.L. (2011). 'Epigenetics and the Placenta', *Human Reproduction Update*, 17: 397–417.

Gene expression in the placenta is regulated by epigenetic processes.

103. Gluckman P.D., Hanson M.D., Cooper C., Thornburg K.L. (2008). 'Effect of In Utero and Early-life Conditions on Adult Health and Disease', *The New England Journal of Medicine*, 359: 61–73. Same first two authors as for the book *Mismatch*.

104. Eriksson J.G., Kajantie E., Osmond C., Thornburg K., Barker D.J.P. (2010). 'Boys Live Dangerously in the Womb', *American Journal of Human Biology*, 22: 330–5.

105. Burrows T.D., King A., Loke Y.W. (1994). 'Expression of Adhesion Molecules by Endovascular Trophoblast and Decidual Endothelial Cells: Implications for Vascular Invasion During Implantation', *Placenta*, 15: 21–3.

106. Loke Y.W., King A. (1995). 'Trophoblast Interaction with Extracellular Matrix', in Loke Y.W., King A. (eds), *Human Implantation: Cell Biology and Immunology*, Cambridge University Press, 151–79.

107. Zhou Y., Damsky C.H., Chiu K., Roberts J.M., Fisher S.J. (1993). 'Pre-eclampsia is Associated with Abnormal Expression of Adhesion Mol-

ecules by Invasive Cytotrophoblasts', *Journal of Clinical Investigation*, 91: 950–60.

108. Jokhi P.P. (1994). 'Cytokines and their Receptors in Human Implantation', PhD thesis (University of Cambridge).

109. Dimitriadis E., White C.A., Jones R.L., Salamonsen L.A. (2005). 'Cytokines, Chemokines and Growth Factors in Endometrium Related to Implantation', *Human Reproduction Update*, 11: 613–30.

110. Seval Y., Korgun E.T., Demir R. (2007). 'Hofbauer Cells in Early Human Placenta: Possible Implications in Vasculogenesis and Angiogenesis', *Placenta*, 28: 841–5. These cells might be the source of angiogenetic growth factors.

CHAPTER 6

111. Loke Y.W. (1978). *Immunology and Immunopathology of the Human Foetal–Maternal Interaction*, Elsevier/North Holland BioMedical Press. This volume describes the early years of research on human reproductive immunology. Our understanding of the subject was rather rudimentary then.

     Loke Y.W., King A. (1995). *Human Implantation: Cell Biology and Immunology*, Cambridge University Press. This second volume shows the progress that had been made by the time we reached the 1990s.

112. Playfair J.H.L., Chain B.M. (2009). *Immunology at a Glance*, Wiley Blackwell.

113. Tauber A.I. (1997). *The Immune Self: Theory or Metaphor?* Cambridge University Press.

114. Medawar P.B. (1953). 'Some Immunological and Endocrinological Problems Raised by the Evolution of Viviparity in Vertebrates', *Symposia of the Society for Experimental Biology*, 7: 320–38. The seminal paper that introduced the concept of the fetus as an allograft.

115. Loke Y.W., King A. (1991). 'Recent Developments in the Human Maternal–Fetal Immune Interaction', *Current Opinion in Immunology*, 3: 762–6.

116. Parham P. (1994). 'The Rise and Fall of Great Class I Genes', *Seminars in Immunology*, 6: 373–82.

117. Hiby S.E., King A., Sharkey A.M., Loke Y.W. (1999). 'Molecular Studies of Trophoblast HLA-G: Polymorphism, Isoforms, Imprinting and Expression in Preimplantation Embryo', *Tissue Antigens*, 53: 1–13.

118. Ryan A.F., Grendell R.L., Geraghty D.E., Golos T.G. (2002). 'A Soluble Isoform of the Rhesus Monkey Nonclassical MHC Class I Molecule

Mamu-AG is Expressed in the Placenta and the Testis', *Journal of Immunology*, 169: 673–83.

119. Caligiuri M.A. (2008). 'Human Natural Killer Cells', *Blood*, 112: 461–9.

120. Moffett-King A. (2002). 'Natural Killer Cells and Pregnancy', *Nature Reviews Immunology*, 2: 656–63.

Trundley A., Moffett A. (2004). 'Human Uterine Leukocytes and Pregnancy', *Tissue Antigens*, 63: 1–12.

121. Croy B.A., Chantakru S., Esadeg S., Ashkar A.A., Wei Q. (2002). 'Decidual Natural Killer Cells: Key Regulators of Placental Development (A Review), *Journal of Reproductive Immunology*, 57: 151–68. A comprehensive review of NK cells in mice.

122. King A., Wellings V., Gardner L., Loke Y.W. (1989). 'Immunocytochemical Characterisation of the Unusual Large Granular Lymphoctyes in Human Endometrium throughout the Menstrual Cycle', *Human Immunology*, 24: 195–205.

123. Kärre K. (2008). 'Natural Killer Cell Recognition of Missing Self', *Nature Immunology*, 9: 477–80.

124. Rajogopalan S., Long E.O. (1999). 'A Human Histocompatibility Leukocyte Antigen (HLA)-G-specific Receptor Expressed on all Natural Killer Cells', *Journal of Experimental Medicine*, 189: 1093–100.

Gómez-Lozano N., de Pablo R., Puente S., Vilches C. (2003). 'Recognition of HLA-G by the NK Cell Receptor KIR2DL4 is Not Essential for Human Reproduction', *European Journal of Immunology*, 33: 639–44.

125. Loke Y.W., King A. (2000). 'Decidual Natural Killer Cell Interaction with Trophoblast: Cytolysis or Cytokine Production?' *Biochemical Society Transactions*, 28: 196–8.

126. Apps R., Gardner L., Moffett A. (2008). 'A Critical Look at HLA-G', *Trends in Immunology*, 29: 313–21. A good review of why the status of HLA-G is so controversial.

127. Chumbley G., King A., Gardner L., Howlett S., Holmes N., Loke Y.W. (1994). 'Generation of an Antibody to HLA-G in Transgenic Mice and Demonstration of the Tissue Reactivity of this Antibody', *Journal of Reproductive Immunology*, 27: 173–86.

128. Zemmour J., Parham P. (1992). 'Distinctive Polymorphism at the HLA-C Locus: Implications for the Expression of HLA-C', *Journal of Experimental Medicine*, 176: 937–50.

129. King A., Burrows T.D., Hiby S.E., Joseph S., Verma S., Lim P.B. et al. (2000). 'Surface Expression of HLA-C Antigen by Human Extravillous Trophoblast, *Placenta*, 21: 376–87.

130. Verma S., King A., Loke Y.W. (1997). 'Expression of Killer-cell Inhibitory Receptors (KIR) on Human Uterine Natural Killer Cells', *European Journal of Immunology*, 27: 979–83.

   Trowsdale J., Moffett A. (2008). 'NK Receptor Interactions with MHC Class I Molecules in Pregnancy', *Seminars in Immunology*, 20: 317–20.

131. Hiby S.E., Walker J.J., O'Shaughnessy K.M., Redman C.W., Carrington M., Trowsdale J., Moffett A. (2004). 'Combinations of Maternal KIR and Fetal HLA-C Genes Influence the Risk of Preeclampsia and Reproductive Success', *Journal of Experimental Medicine*, 200: 957–65.

132. Hiby S.E., Apps R., Sharkey A.M., Farrell L.E., Gardner L., Mulder C. et al. (2010). 'Maternal Activating KIRs Protect Against Human Reproductive Failure Mediated by Fetal HLA-C2', *Journal of Clinical Investigation*, 120: 4102–10. The authors are now in the process of identifying which activating KIR is responsible for the protection.

133. Parham P. (2005). 'MHC Class I Molecules and KIRs in Human History, Health and Survival', *Nature Reviews Immunology*, 5: 201–14. A comprehensive review covering all aspects of KIRs.

134. Colucci F., Boulenouar S., Kieckbusch J., Moffett A. (2011). 'How Does Variability of Immune System Genes Affect Placentation?' *Placenta*, 32: 539–45.

   Lash G.E., Bulmer J.N. (2011). 'Do Uterine Natural Killer (uNK) Cells Contribute to Female Reproductive Disorders?' *Journal of Reproductive Immunology*, 88: 156–64.

135. Golos T.G., Bondarenko G.I., Dambaeva S.V., Breburda E.E., Durning M. (2010). 'On the Role of Placental Major Histocompatibility Complex and Decidual Leukocytes in Implantation and Pregnancy Success Using Non-human Primate Models', *International Journal of Developmental Biology*, 54: 431–43.

   Donaldson W.L., Oriol J.G., Pelkaus C.L., Antczak D.F. (1994). 'Paternal and Maternal Major Histocompatibility Complex Class I Antigens are Expressed Co-dominantly by Equine Trophoblast', *Placenta*, 15: 123–35. These are the invasive cells of the chorionic girdle that form the endometrial cups. Other trophoblast populations do not express MHC antigens. An analogous situation is seen in humans.

Noronha L.E., Huggler K.E., de Mestre A.M., Miller D.C., Antczak D.F. (2012). 'Molecular Evidence for Natural Killer-like Cells in Equine Endometrial Cups', *Placenta*, 33: 379–86. Remarkable conservation of NK cells since horses and humans diverged 100 million years ago. Cup cells also express MHC antigens so dialogue between trophoblast MHC and uterine NK cells seems to be important for pregnancy, as seen also in humans. But the horse uterus does not show spiral artery alteration as seen in species with haemochorial placentation. So what is the function of this MHC–NK cell interaction in the horse?

CHAPTER 7

136. Hinson J., Raven P., Chew S. (eds) (2007). *The Endocrine System: Basic Science and Clinical Conditions*, Churchhill Livingston, Elsevier.
137. Villee D.B. (1969). 'Development of Endocrine Function in the Human Placenta and Fetus', *New England Journal of Medicine*, 281: 473–84.
138. Dreskin R.B., Spicer S.S., Greene W.B. (1970). 'Ultrastructural Localization of Chorionic Gonadotropin in Human Term Placenta', *Journal of Histochemistry and Cytochemistry*, 18: 864–73.

    Fox H., Kharkongor F.N. (1970). 'Immunofluorescent Localization of Chorionic Gonadotropin in the Placenta and in Tissue Cultures of Human Trophoblast', *Journal of Pathology*, 101: 277–82.
139. Cooke L., Nelson S.M. (2011). 'Reproductive Ageing and Fertility in an Ageing Population: A Review', *The Obstetrician & Gynaecologist*, 13: 161–8.
140. Hussa R.O. (1980). 'Biosynthesis of Human Chorionic Gonadotropin', *Endocrine Review*, 1: 268–94.
141. Loke Y.W., Wilson D.V., Borland R. (1972). 'Localization of Human Gonadotropin in Monolayer Cultures of Trophoblast Cells by Mixed Agglutination', *American Journal of Obstetrics & Gynecology*, 133: 875–9. This technique demonstrates that HCG is on the surface of trophoblast cells.
142. Cole L.A. (2010). 'Hyperglycosylated HCG, a Review', *Placenta*, 31: 653–64.
143. Pattillo R.A., Hussa R.O., Yorde D.E., Cole L.A. (1983). 'Hormone Synthesis by Normal and Neoplastic Human Trophoblast', in Loke Y.W., Whyte A. (eds), *Biology of Trophoblast*, Elsevier, 283–316.
144. Braunstein G.D., Vaitukaitis J.L., Carbone P.P., Ross G.T. (1973). 'Ectopic Production of Human Chorionic Gonadotropin by Neoplasms', *Annals of Internal Medicine*, 78: 39–45. There is an extensive literature documenting the presence of HCG in the serum of patients with non-trophoblast

neoplasms. It is not clear why gonadotrophin-producing tumours tend to secrete HCG but not LH or FSH.

145. Vaitukaitis J.L. (1975). 'Placental Proteins and their Subunits as Tumour Markers', *Annals of Internal Medicine*, 82: 71.

146. Fowden A.L., Forhead A.J. (2009). 'Hormones as Epigenetic Signals in Developmental Programming', *Experimental Physiology*, 94: 607–25.

147. Ada G.L., Basten A., Jones W.R. (1985). 'Prospects for Developing Vaccines to Control Fertility', *Nature*, 317: 288–9.

148. Swaminathan H., Bahl O.P. (1970). 'Dissociation and Recombination of the Subunits of Human Chorionic Gonadotropin', *Biochemical and Biophysical Research Communications*, 40: 422–7.

   Fiddes J.C., Goodman H.M. (1980). 'The cDNA for the Beta-subunit of Human Chorionic Gonadotropin Suggests Evolution of a Gene by Readthrough into the 3'-Untranslated Region', *Nature*, 286: 684–7. The authors deduced that the 30 amino acid carboxyl-terminal extremeties of HCG beta may have arisen by the loss of the termination codon of an ancestral beta-like gene.

   Henke A., Gromoll J. (2008). 'New Insights into the Evolution of Chorionic Gonadotropin', *Molecular and Cellular Endocrinology*, 291: 11–19.

149. Stevens V.C. (1986). 'Current Status of Antifertility Vaccines Using Gonadotropin Immunogens', *Immunology Today*, 7: 369–74.

150. Josimovich J.E., MacLaren J.A. (1962). 'Presence in the Human Placenta and Term Serum of a Highly Lactogenic Substance Immunologically Related to Pituitary Growth Hormone', *Endocrinology*, 71: 209–20. This observation led to the isolation and characterization of HPL.

151. Haig D. (1996). 'Placental Hormones, Genomic Imprinting and Maternal–Fetal communication', *Journal of Evolutionary Biology*, 9: 357–80.

152. Ha C., Waterhouse R., Wessells J., Wu J., Dvcksler G.S. (2005). 'Binding of Pregnancy-specific Glycoprotein 17 to CD9 on Macrophages Induces Secretion of IL-10, IL-6, PGE_2, and TGF-B,', *Journal of Leukocyte Biology*, 77: 948–56. Receptor CD9 is in the mouse. The human receptors for PSGs are not known. But the authors suggest that binding to different receptors can still use similar signalling mechanisms to lead to secretion of the same cytokines.

   Wynne F., Ball M., McLellan A.S., Dockery P., Zimmermann W., Moore T. (2006). 'Mouse Pregnancy-specific Glycoproteins: Tissue-specific Expression and Evidence of Association with Maternal Vasculature', *Reproduction*, 131: 721–32.

153. Bonnin A., Goeden N., Chen K., Wilson M.L., King J., Shih J.C. et al. (2011). 'A Transient Placental Source of Seratonin for the Fetal Forebrain', *Nature*, 472: 347–50.

154. Zeltser L.H., Leibel R.L. (2011). 'Roles of the Placenta in Fetal Brain Development', *PNAS USA*, 108: 15667–8.

CHAPTER 8

155. Jansson T., Powell T.L. (2000). 'Placental Nutrient Transfer and Fetal Growth', *Nutrition*, 16: 500–2.

    Jones H.N., Powell T.L., Jansson T. (2007). 'Regulation of Placental Nutrient Transport: A Review', *Placenta*, 28: 763–74.

156. Mayhew T.M. (2009). 'A Stereological Perspective on Placental Morphology in Normal and Complicated Pregnancies', *Journal of Anatomy*, 215: 77–90.

157. Gadsby R., Barnie-Adshead T. (2011). 'Severe Nausea and Vomiting of Pregnancy: Should it be Treated with Appropriate Pharmacotherapy? A Review', *The Obstetrician & Gynaecologist*, 13: 107–11. There are over 25,000 admissions per year for hyperemesis gravidarium in England.

158. Jansson T. (2001). 'Amino Acid Transporters in the Human Placenta', *Pediatric Research*, 49: 141–7.

159. Jansson T., Wennergren M., Powell T.L. (1999). 'Placental Glucose Transport and GLUT 1 Expression in Insulin-dependent Diabetes', *American Journal of Obstetrics & Gynecology*, 180: 163–8.

160. Jansson T., Powell T.L. (2006). 'Human Placental Transport in Altered Fetal Growth: Does the Placenta Function as a Nutrient Sensor?—A Review', *Placenta*, 27, Suppl. A: S91–7.

161. Loke Y.W. (1978). 'Transmission of Immunoglobulins from Mother to Foetus Before Birth', in Loke Y.W., *Immunology and Immunopathology of the Human Foetal–Maternal Interaction*, Elsevier/North Holland BioMedical Press, 119–39. This volume has an entire chapter on the subject. It reviews the evidence that antibody transfer across the placenta is dependent on antibody class and not antibody size.

162. Brambell F.W.R. (1970). *The Transmission of Passive Immunity from Mother to Young*, North Holland. A comprehensive monograph devoted to the subject.

163. Firan M., Bawdon R., Radu C., Ober R.J., Eaken D., Antohe F., Ghetie V., Ward E.S. (2001). 'The MHC Class I-related Receptor FcRn Plays an Es-

sential Role in the Maternofetal Transfer of Gamma-globulin in Humans', *International Immunology*, 13: 993–1002.

164. Roopenian D.C., Akilesh S. (2007). 'FcRn: The Neonatal Fc Receptor Comes of Age', *Nature Reviews Immunology*, 7: 715–24.

165. Mollison P.L. (1973). 'Clinical Aspects of Rh Immunisation', *American Journal of Clinical Pathology*, 60: 287–301.

166. Swinburne L.M. (1970). 'Leucocyte Antigens and Placental Sponge', *Lancet*, 2: 592–4.

     Szulman A.E. (1972). 'The A, B and H Blood Group Antigens in Human Placenta', *New England Journal of Medicine*, 286: 1028–31.

167. Clarke C.A. (1968). 'Prevention of Rhesus Iso-immunisation', *Lancet*, 2: 1–7.

168. Whitley C.B., Kramer K. (2010). 'A New Explanation for the Reproductive Woes and Midlife Decline of Henry VIII', *The Historical Journal*, 53: 827–48.

169. Serr D.M., Ismajovich B. (1963). 'Determination of the Primary Sex Ratio from Human Abortions', *American Journal of Obstetrics & Gynecology*, 87: 63–5.

170. Borland R., Loke Y.W., Oldershaw P.J. (1970). 'Sex Difference in Trophoblast Behaviour on Transplantation', *Nature*, 228: 572. In experiments with guinea pigs, allografts from female trophoblast survived considerably longer than from male trophoblast.

171. Hadley A.G., Soothill P. (eds) (2002). *Autoimmune Disorder in Pregnancy*, Cambridge University Press.

172. Nelson J.L. (1996). 'Maternal-fetal Immunology and Autoimmune Disease: Is Some Autoimmune Disease Auto-alloimmune or Allo-autoimmune?' *Arthritis & Rheumatism*, 39: 191–4.

173. Loke Y.W. (1978). 'Transmission of Immunity after Birth', in Loke Y.W. *Immunology and Immunopathology of the Human Foetal–Maternal Interaction*, Elsevier/North Holland BioMedical Press, 211–21. This volume has an entire chapter on the subject.

174. Loke Y.W., Eremin O., Ashby J., Day S. (1982) 'Characterization of the Phagocytic Cells Isolated from the Human Placenta', *Journal of the Reticuloendothel Society*, 31: 317–24.

175. Silverstein A.M. (1972). 'Fetal Immune Responses in Congenital Infection', *New England Journal of Medicine*, 286: 1413–14.

176. Cederqvist L.L., Kimball A.C., Ewool L.C., Litwin S.D. (1977). 'Fetal Immune Response Following Congenital Toxoplasmosis', *Obstetrics &*

Gynecology, 50: 200–4. Reports that infants with congenital toxoplasmosis all had elevated serum IgM but children born to mothers with toxoplasmosis who themselves were not infected did not have raised IgM levels. This shows that the fetus itself is capable of making antibodies. But this ability does not reach full maturity until several months after birth.

177. Cramer D.V., Kunz H.W., Gill T.J. (1974). 'Immunologic Sensitization Prior to Birth', *American Journal of Obstetrics & Gynecology*, 120: 431–9.

178. Ferguson-Smith M.A. (2003). 'Placental mRNA in Maternal Plasma: Prospects for Fetal Screening', *PNAS USA*, 100: 4360–2.

Bianchi D.W. (2012). 'Fetal Genes in Mother's Blood', *Nature*, 487: 304–5.

179. Bianchi D.W. (2000). 'Fetomaternal Cell Trafficking: A New Cause of Disease?' *American Journal of Medical Genetics*, 91: 22–8.

180. Isoda T., Ford A.M., Tomizawa D., van Delft F.W., De Castro D.G., Norkio M. et al. (2009). 'Immunologically Silent Cancer Clone Transmission from Mother to Offspring', *PNAS USA*, 106: 17882–5.

181. Attwood H.D., Park W.W. (1961). 'Embolism to the Lungs by Trophoblast', *Journal of Obstetrics & Gynaecology of the British Commonwealth*, 68: 611–17.

Douglas G.W., Thomas L., Carr M., Cullen N.M., Morris R. (1959). 'Trophoblast in the Circulating Blood During Pregnancy', *American Journal of Obstetrics & Gynecology*, 78: 960–73.

CHAPTER 9

182. Boyd J.D., Hamilton W.J. (1970). 'Historical Survey', in *The Human Placenta*, Cambridge, W. Heffer & Sons.

Steven D.H. (ed.) (1975). *Comparative Placentation: Essays in Structure and Function*, Academic Press. Donald Steven's own chapter, 'Placenta Depicta—An Illustration and Ideas', 1–24, is an interesting account of the placenta through the ages from ancient Egypt to the 20th century. Beautiful illustrations.

183. Wilms S. (1992). 'Childbirth Customs in Early China', an MA thesis submitted to the Department of East Asian Studies, University of Arizona, USA. Based on the discovery (1970s) of Chinese medical manuscripts in the tomb of Ma-Wang-Tui in Chang-Sha dated to the 2nd century. Constitutes one of the earliest available collections of medical documents.

184. Li Shizhen (1518–93) *Compendiium of Materia Medica*. The author spent 30 years compiling this magnum opus. The Compendium covers the

period between remote antiquity and the Ming Dynasty when the Compendium was published in 1593.

185. Prusa A.R., Marton E., Rosner M., Bernaschek G., Hengstschläger M. (2003). 'Oct-4-expressing Cells in Human Amniotic Fluid: A New Source for Stem Cell Research?' *Human Reproduction*, 18: 1489–93.

Cananzi M., Atala A., De Coppi P. (2009). 'Stem Cells Derived from Amniotic Fluid: New Potentials in Reproductive Medicine', *Reproductive BioMedicine Online*, 18: 17–27.

186. Young S.M., Benyshek D.C. (2010). 'In Search of Human Placentophagy: A Cross-cultural Survey of Human Placenta Consumption, Disposal Practices and Cultural Beliefs', *Ecology of Food and Nutrition*, 49: 467–84. This study surveyed 179 human societies. Placentophagy is rare in prehistoric, historic, and contemporary human cultures.

Young S.M., Benyshek D.C., Lienard P. (2012). 'The Conspicuous Absence of Placenta Consumption in Human Postpartum Females: The Fire Hypothesis', *Ecology of Food and Nutrition*, 51: 198–217. The thought that the placenta might harbour toxic chemicals derived from the mother's exposure to smoke and ash acts as a deterrent to placentophagy.

187. Barker D.J., Thornburg K.L., Osmond C., Kajantie E., Eriksson J.G. (2009). 'The Surface Area of the Placenta and Hypertension in the Offspring in Later Life', *International Journal of Developmental Biology*, 54: 525–30.

Egbor M., Ansari T., Morris N., Green C.J., Sibbons P.D. (2006). 'Pre-eclampsia and Fetal Growth Restriction: How Morphometrically Different is the Placenta?' *Placenta*, 27: 727–34.

188. *Placenta*. Official Journal of the International Federation of Placenta Associations (IFPA). Published monthly by Elsevier. Available online at SciVerse ScienceDirect. For those readers who are interested in getting the latest news about the placenta, this is the journal to read. It publishes cutting-edge research papers on all aspects of the placenta, human as well as other mammalian species. In addition, there are regular review articles which the general reader will find useful.

Index